# Spontaneous Alternation Behavior

W.N. Dember C.L. Richman

# Spontaneous Alternation Behavior

Springer-Verlag
New York Heidelberg Berlin
London Paris Tokyo Hong Kong

William N. Dember
Department of Psychology
University of Cincinnati
Cincinnati, OH 45221-0376, USA

Charles L. Richman
Department of Psychology
Wake Forest University
Winston-Salem, NC 27109, USA

Library of Congress Cataloging-in-Publication Data
Spontaneous alternation behavior / edited by William N. Dember and Charles L. Richman ; contributors, William N. Dember . . . [et al.].
p. cm.
Includes bibliographical references.
ISBN-13: 978-1-4613-8881-4
1. Animal psychology. 2. Animal behavior. I. Dember, William N. (William Norton), 1928– II. Richman, Charles L.
QL785.S74 1989
591.51–dc20 89-19680

Printed on acid-free paper.


Softcover reprint of the hardcover 1st edition 1989



Typeset by Publishers Service, Bozeman, Montana.

9 8 7 6 5 4 3 2 1

ISBN-13: 978-1-4613-8881-4 e-ISBN-13: 978-1-4613-8879-1
DOI: 10.1007/978-1-4613-8879-1

This book is dedicated to Joyce Richman
and the Richman children, Susan and Adam
and to
Cynthia Dember and the Dember children,
Joanna Engelke, Laura, and Greg

# Preface

A wide variety of species, including human beings, exhibits a remarkably reliable behavior pattern, known as spontaneous alternation behavior (SAB), that has intrigued researchers for over seven decades. Though the details may vary depending on species and setting, SAB essentially entails first choosing one member of a pair of alternatives and then the other, without instructions or incentives to do so. Spontaneous alternation is manifested even in the early trials of a discrimination-learning experiment, where only one of the choices is reinforced. Indeed, that was the setting in which SAB was first noted (Hunter, 1914). Reinforcement contingencies, evidently, are superimposed, not on a random sequence of choices, but on a potent, systematic behavior pattern.

This book is the first to be devoted entirely to SAB and closely related phenomena, such as habituation and exploration. The literature on SAB is vast, covering a host of questions ranging from the cues that guide alternation to its phylogenetic and ontogenetic generality, its relation to learning and motivation, and its neurochemical substrates. In separate chapters we take up each of the major issues, reviewing what is known about the several facets of SAB and revealing areas of ignorance. The chapter authors were encouraged to discuss their own research where pertinent, some of it as yet unpublished, indeed some conducted specifically for this volume.

While the chapters are ostensibly devoted to different aspects of SAB research, there is inevitable overlap. For example, one cannot discuss brain structures implicated in SAB without also referring to neural transmitters; by the same token, a discussion of drug effects on SAB would be incomplete without reference to the sites in the brain where these drugs make impact. In a similar vein, the ontogeny of SAB is at least in part attributable to the maturation of specific brain centers, and research on neural structures subserving SAB can be informed by data from ontogenetic studies. The contributing authors, while aware of the topics to be covered in the various chapters, prepared their own chapters independently of one another. It is interesting to see, in light of that, how the same data can sometimes be given a different slant by authors with somewhat different missions or orientations. The editors, of course, tried to minimize outright contradictions of fact across chapters, but in general we allowed the contributors the freedom to interpret the literature in a way that best fit their own conceptions. Despite over seven decades of research, we are well aware that most assertions about SAB ought to be treated as tentative hypotheses rather than as incontrovertible conclusions.

The editors conceived the idea of a book devoted to SAB after preparing a "review of the literature" for journal publication (Richman, Dember, & Kim, 1987). We found the literature since the last comprehensive review (Dember & Fowler, 1958) to be so extensive that we could not fully do it justice even in a lengthy article. Moreover, we were aware of technical issues that were better addressed by specialists. Our first step was to organize a symposium on SAB and exploratory behavior for the 1987 meeting of the American Psychological Association. Most of the symposium participants agreed to develop their talks into chapters for this volume. We were also fortunate to have been able to add to the list of contributing authors Norman Spear and James Miller, who have written a concise but exciting chapter on the ontogeny of SAB. Also solicited for the book was a chapter by Robert Douglas on brain mechanisms subserving SAB; Douglas also shares his wisdom, in a concluding chapter, on some very practical problems in conducting SAB research, the theme of his symposium presentation.

While originally the province of animal psychologists, SAB has proven of interest to researchers in several of psychology's subdisciplines, as well as to researchers in related fields. The bulk of recent work, in fact, has been performed by neural scientists using SAB as a simple, convenient, reliable, and relatively "noninvasive" indicator response rather than as an object of investigation in its own right. Of course, there are still great gaps in our fundamental knowledge of SAB. We expect, therefore, that psychologists, along with other biobehavioral scientists and students, will find something of value in this volume for their own work, whether basic or applied.

The contributors to this book come from disparate backgrounds; all, however, directly or indirectly, are indebted to Edward L. Walker, who played a seminal role in the modern era of SAB research as theorist and experimenter as well as mentor or "grandmentor" to many of those whose work is herein represented. We would also like to acknowledge the invaluable assistance of Mrs. Shirley Doxsey, who typed the very lengthy bibliography and indexes. Finally, we offer a special note of thanks to Joanna Dember Engelke, who, a decade ago, devoted winter break of her freshman year in college to conducting the initial literature search (see Dember & Dember, 1983) that paved the way for the review article that led to the symposium that begat the writing of this book.

William N. Dember
Charles L. Richman

# Contents

# Contributors


William N. Dember
Department of Psychology
University of Cincinnati
Cincinnati, OH 45221, USA

Robert J. Douglas
Department of Psychology
University of Washington
Seattle, WA 98195, USA

Robert N. Hughes
Department of Psychology
University of Canterbury
Christchurch 1, New Zealand

Larry Kokkinidis
Department of Psychology
University of Saskatchewan
Saskatoon, Canada S7N OWO

James S. Miller
Department of Psychology
State University of
New York, Binghamton
Binghamton, NY 13901, USA

Charles L. Richman
Department of Psychology
Wake Forest University
Winston-Salem, NC 27109, USA

Norman E. Spear
Department of Psychology
State University of
New York, Binghamton
Binghamton, NY 13901, USA

# 1
# Historical Overview

WILLIAM N. DEMBER
UNIVERSITY OF CINCINNATI

*What is SAB?*

The focus of this chapter, and of this book as a whole, is a simple, robust, near-universal response pattern dubbed spontaneous alternation behavior. This pattern of responding can be described in a variety of ways. For our purposes, the following quote quite adequately captures the essence of the phenomenon:

> A rat is placed in the start stem of a T-maze and given two trials with a brief intertrial interval. On trial 1 ($t_1$) it enters one of the goal arms; gently removed from the maze and then returned for its second trial, the animal on trial 2 ($t_2$) enters the other goal arm. This pattern of entering first one arm and then the alternate arm, if exhibited with significant frequency by a given animal over multiple testings or on tests of many animals, has been called spontaneous alternation behavior (SAB) – alternation for obvious reasons and spontaneous because the animal has not been trained through differential reinforcement to behave that way. (Richman et al., 1987, p. 358)

What else might an animal do in this situation other than alternate? There seem to be two possible scenarios: (1) Assuming that what a rat does under a given set of circumstances is the best predictor of what it will do when again faced with those circumstances, one might with confidence expect the behavior on $t_2$ to be the same as that on $t_1$ (i.e., repetition). Thus, if an animal turned into goal arm A on $t_1$, it might reasonably be expected to turn into A on the next occasion if it is given the opportunity to make that choice. The prediction of repetition rests on the assumption that the animal's behavior on $t_1$ reflected a real choice and was not merely the outcome of some internal random generator – that its entering A on $t_1$ was not analogous to a flipped coin's coming up heads. (2) The response on $t_1$ might best be conceived as "random," or at least as the outcome of processes that are unlikely exactly to recur at the time of $t_2$. Thus, behaviors on $t_1$ and $t_2$ will be independent of one another. Note that the first prediction, repetition, and the second, independence, converge if the probability of turning A on any trial is 1.00 – that is, if there is such a strong bias in favor of A that B is never chosen. If there is no bias at all, if the probability of entering A is 0.50, then the probability of the alternation pattern (A followed by B or B followed by A) is also 0.50. One problem for researchers has been to determine a proper "chance" baseline probability against which statistically to evaluate obtained values of alternation probability. A bit more on that issue later.

As already indicated, of the three possible scenarios, it is spontaneous alternation behavior (abbreviated henceforth as SAB) that typically prevails. Since that pattern was first reported, hundreds of articles have been published with SAB as their concern. The experiments described in those articles have covered the full range of variables and processes dear to the hearts of psychologists, from manipulations of maze stimuli to pharmacological and neurobiological interventions. Those efforts, at their core, have been aimed, directly or indirectly, at finding out why animals alternate. It turns out that SAB, perhaps the most reliable phenomenon in all of psychological research, is not nearly as simple as it might seem; while enormous progress has been made over the past half century in elucidating pertinent variables and likely underlying mechanisms, a completely satisfying answer to the fundamental question, why alternation, remains elusive.

The succeeding chapters will discuss in detail the major lines of research on SAB. This chapter is intended to set the stage for those discussions by providing a historical overview of research and speculation on this intriguing phenomenon. Issues will be introduced in this chapter and then more thoroughly examined in subsequent chapters. The orientation here will be mainly historical, emphasizing the period from the 1920s through the 60s.

## *Early Observations and Speculations*

### *Pioneering Studies*

Establishing scientific primacy is always difficult. In the current instance, credit for observing and reporting the alternation tendency probably goes to Walter Hunter (1914), who in an article on auditory sensitivity in the rat, using a two-alternative discrimination-learning paradigm, remarked in a footnote: "A rat may alternate between sides in the order right-left. If this leads to severe punishment, he may at times reverse the alternation to a left-right order." Hunter also describes the behavior of a particular animal (rat number 5) that "formed a habit of alternating after *each success only* (author's italics). He would go to the right and, if successful, would go to the left in the next trial. Had he failed on the right, though, he would have continued to go there until he succeeded" (Hunter, 1914, p. 216). In modern terminology, rat number 5's behavior would be characterized as reflecting a "win-shift strategy," one of several explanatory concepts for SAB to be offered over the next half century (see Chapters 2 and 4 for further discussion). Ten years later, Edward C. Tolman also noted the alternation tendency and brought it to the attention of others in an article entitled "Purpose and Cognition: The Determiners of Animal Learning." In that article, which deals primarily with the issue of how to conceptualize animal learning without recourse "to the customary physiological notions of stimulus, neural excitation, synaptic resistance, and muscle contraction" (Tolman, 1925, p. 285), Tolman remarks, almost in passing, on an interesting observation of rats' behavior in a T maze so arranged that both goal arms led to food reward:

> . . . although some animals acquired a consistent preference for one side rather than for the other, this preference often appeared relatively late and in some animals not at all and as long as such a preference was absent there was instead a very pronounced tendency toward continuing and regular alternation – left, right, left, right, or right, left, right, left. That is, not only did the animals tend to average five times to the right and five times to the left, out of the total of ten trials on any day, but the percentages of immediate alternations from left to right and right to left were much greater than would have resulted from mere chance. It appeared, in short, that even where either side was equally "satisfactory" there was in our rats a positive tendency left over toward variation of response. (Tolman, 1925, pp. 290–291)

Tolman does little more with that observation than to cite it as evidence of some sort of exploratory tendency manifested early in the course of maze learning. Tolmaniacs, if there are any, should find that article interesting for what it reveals about the origins of purposive behaviorism (Tolman, 1932).

SAB was rescued from obscurity by Wayne Dennis, who initially was not aware of Tolman's observations (Dennis, 1939, p. 305). Dennis had been investigating maze learning in rats and had accepted the implicit assumption shared by learning researchers that before the application of differential reinforcement became effective, the animal's maze behavior was essentially random. That is, learning could be thought of as imposing predictability on behavior that initially was haphazard. But the rat's initial choices among maze paths were systematic, not random, following the pattern characterized as SAB.

### *The Basic Paradigm*

Dennis's earliest studies (Dennis & Henneman, 1932; Dennis, Henneman, & Varner, 1932) employed multiple-unit, elevated mazes. Because of the nature of those mazes, there might have been an excessively long interval between successive opportunities to enter the choice arms of any given maze unit, thereby obscuring the full strength of the SAB tendency. Correctly assuming that the alternation pattern would better be revealed if the time interval between "trials" were shortened, Dennis followed up those initial experiments with studies that employed only a single-unit Y maze (Dennis, 1935; Dennis & Sollenberger, 1934). That apparatus change permitted a brief intertrial interval. The procedure used by Dennis provides the paradigm for most subsequent research. Specifically, "the experimenter, as soon as the subject made a choice, picks it up and returns it to the original alley, where it remains until a second choice is made" (Dennis, 1935, p. 488). In prior research using multiple-unit mazes, Dennis notes that "the animal frequently *brought itself* back to the choice-point" (Dennis, 1935, p. 488, italics added). I emphasize the difference in procedure here for two reasons: (1) Some researchers have found it preferable to follow the original method of allowing the animal to move freely throughout the maze (whether of the single- or multiple-unit variety), whereas others have adopted the modified procedure, which requires experimenter intervention between trials. Though it is not clear that this procedural variation is an important one, cross-study comparisons may be muddied by such differences. (2) The experimenter-intervention procedure suggests a potential explanation of SAB – namely, that being abruptly picked up and removed from a goal arm by a strange giant (the human experimenter) may be aversive to rats, and SAB may simply reflect one-trial avoidance learning. There are many reasons for rejecting this attractive explanation, chief among them the fact that SAB is evident when the other procedure is followed. And to be fair to the modified procedure, the animals in the typical SAB study are thoroughly tamed and familiar with the experimenter; at the same time, good experimenters know how to handle their animals in a gentle, nonaversive manner. Finally, as was pointed out by Glenn Heathers (1940), those animals least disturbed by handling show the greatest level of SAB, while "rats which frightened easily were very apt to fixate one pathway, hence alternate less than the average" (Heathers, 1940, p. 374).

Despite the historical order in which the two procedures for getting the animal ready to launch $t_2$ were introduced, we will adopt the convention of treating the experimenter-intervention method as the paradigmatic SAB procedure, with other methods considered variants.

Along these same lines, note that Dennis used an elevated maze; the pathways, or "arms" of such a maze are open platforms on which the animal locomotes; what keeps the animal from leaving the maze is the fact that it is raised off the floor at a sufficient height to discourage defection. Subsequent researchers typically used enclosed mazes – that is, mazes having alleys with sides and tops that prevented the animal from leaving the confines of the maze. Whether that difference in maze structure matters is also not clear, though there are data to suggest it may not. Specifically, in his research using the elevated maze Dennis (1935) reports an alternation rate of about 0.82: SAB is evident on about five of six tests. That same rate comes up over and over again in subsequent research using enclosed mazes.

Without now going into the myriad methodological differences that abound in the SAB literature (e.g., Is the maze floor solid or wire mesh? How are the interiors of the arms of an enclosed maze decorated? Are the arms the same, or do they differ in brightness or pattern? What are the dimensions of the maze? Are the maze arms cleaned between trials? Are the animals hungry or sated? How many trials are run per day, with what intertrial interval), it is sufficient to note that there are lots of ways to run an SAB experiment. The outcome may be very sensitive to procedural detail, a bit of a nuisance, of course, but also strong indication that the apparent simplicity of SAB can be quite misleading. As with most other behavioral phenomena, the more closely we examine it, the more puzzling and intriguing it becomes.

One more procedural point before we move on: The single unit of Dennis's maze had the configuration of a Y. A great deal of subsequent research was conducted in mazes that had a T-shaped configuration. That difference can matter, as will be discussed in the next chapter, not globally, in terms of overall alternation rate, but rather in more subtle ways. Similar subtle differences may apply to the distinction between elevated and enclosed mazes, though here, as far as I know, there are no pertinent data from which to draw a conclusion.

### *SAB and Exploration*

The research by Dennis and collaborators established the reality of the alternation pattern. What about interpretation? From the language that both Tolman and Dennis employed in describing the phenomenon, it seems certain that they both viewed it as a manifestation of a more general exploratory tendency. Tolman is quite explicit. He writes, just before giving the description of alternation quoted earlier, "How fast and in what manner a rat 'learns' will be conditioned among other things by the range, methodicalness, and flexibility of his exploratory impulses" (1925, p. 290). A bit later, Tolman raises some empirical questions that have been echoed in subsequent research (e.g., DeValois, 1954; Fowler, Fowler, & Dember, 1959).

> The questions concerning the causal determiners of these initial exploratory propensities is [sic] still unanswered. How much are they due to experience? How much are they conditioned by the particular type of goal and the particular type of physiological need back of that goal? Do, for example, fear and sex call out different initial exploratory tendencies from hunger and from one another, and so on? And how much, finally, do they vary in kind or extensity with the mere urgency and intensity of the need? Would we find that the more intense the motive, the more restricted the exploratory readiness, and the less intense the motive, the more leisurely and widespread the sallies? (Tolman, 1925, p. 291)

Some of these questions are directly addressed in Chapter 4 of this volume.

Dennis also uses the term "explore" or its variants, e.g., "Each rat was given an opportunity to explore the maze" (Dennis & Henneman, 1932, p. 398). "All of the rats which explored behaved in a non-random fashion . . ." (p. 403). It is not quite clear, however, that Dennis intended to make a theoretical statement or was simply using a convenient, and at that time conventional, verb for communicating his observations. In a similar vein, Dennis refers to the rat's "choices" among alternative maze alleys, but again may not necessarily be attributing cognitions to his animals. It is difficult to describe behavior without using words that may have surplus theoretical connotations. To pursue this a bit further, Dennis also typically has his rats "entering" alleys. The point is that he does not refer to animals' making turning responses, such as "turning right" or "turning left." That kind of terminology entered the SAB researcher's lexicon a decade later under the influence of Clark Hull's behavior theory and helped generate, or at least highlight, two very distinct ways of understanding SAB. How you say it undoubtedly can determine how you see it, and, of course, vice versa.

Before those theoretical issues became crystalized, however, additional empirical research was conducted, much of it concerned with temporal parameters. Recall that Dennis switched from complex mazes to the simple Y maze in order to gain better control over the interval between successive exposures to the same choice-point (Dennis, 1935; Dennis & Sollenberger, 1934). Several studies ensued in which the length of the intertrial interval (ITI) was systematically varied and its effect on probability of SAB noted (e.g., Dennis, 1939; Heathers, 1940). In general, the longer the intertrial interval, the lower the probability of SAB. That finding, it turned out, fit very nicely with an account of alternation derived from Hull's behavior theory (Hull, 1943), which located the source of SAB in the animal's response system rather than in some hypothetical exploratory impulse.

## *SAB in the Forties: The Hullian Heyday*

### *Hull's Behavior Theory*

Clark Hull drew on diverse sources–Thorndike, Watson, Pavlov, Darwin–to develop an ambitious, formal ("hypothetico-deductive") behavior theory that came to dominate American psychology in the 1940s and to a lesser extent in the 50s. His general behavior theory was built on a theory of learning, emphasizing the way in which habits (stimulus-response bonds) are established, expressed, and extinguished. It is the last process, extinction, that provided a plausible, though serendipitous account of SAB. Hull postulated two forms of inhibition implicated in the extinction process, one temporary and the other permanent. The former was referred to as reactive inhibition, symbolized as $I_R$, the latter as conditioned inhibition, symbolized as $_SI_R$. Though our current interest lies primarily in $I_R$, a brief discussion of Hull's conceptions of learning and extinction is necessary at this point.

For Hull, learning consisted of the strengthening of stimulus-response bonds. An increment accrued to habit strength ($_SH_R$) whenever a stimulus, S, and a response, R, occurring in close temporal proximity (with S preceding R) were followed immediately (within half a second) by a reduction in some drive. The greater the number of such reinforced S-R pairings, or "training trials," the more closely the value of $_SH_R$ approached the maximum possible value under the training circumstances. Once established, habits were essentially permanent–an assumption Hull shared with Pavlov. How then did he account for "broken habits," that is, for experimental extinction? The procedure for producing extinction–the failure to perform R in the presence of S–is clear: elicit R in the presence of S, but do not follow R with reinforcement. After enough such extinction trials, S will no longer evoke R, not, according to Hull, because the habit itself has been weakened, but rather because of the development of inhibition.

Specifically, in Hull's theory inhibition subtracts from the tendency to make R in the presence of S, a tendency ("excitatory potential," or $_SE_R$) determined by the multiplicative interaction between $_SH_R$ and drive level (D). That is $_SH_R \times D = {}_SE_R$. Inhibition then subtracts from $_SE_R$ to yield "effective excitatory potential," $_S\bar{E}_R$.

As noted, Hull postulates two forms of inhibition, $I_R$ and $_SI_R$. Those two summate to yield the momentary value of net inhibition ($\dot{I}_R$). That is, $\dot{I}_R = I_R + {}_SI_R$.

It is $\dot{I}_R$ that subtracts from ${}_SE_R$ to yield ${}_S\bar{E}_R$ : ${}_S\bar{E}_R = {}_SE_R - \dot{I}_R$. It is from ${}_S\bar{E}_R$, "effective reaction potential," that behavior ultimately flows.

Why both a temporary and a permanent form of inhibition? The need to postulate temporary inhibition ($I_R$) derives from the phenomenon known as spontaneous recovery. If, in a classical conditioning paradigm, an extinction series is carried out on one day to the point where the conditioned stimulus (CS) reliably fails to elicit the conditioned response (CR), and then the animal is tested on the next day, it is often observed that the CS will still evoke the CR on the early trials of that day's extinction series. This reappearance of the CR on one day following apparently complete extinction on the preceding day was called spontaneous recovery by Pavlov. The inference from spontaneous recovery is that a portion of the extinction on a given day is attributable to a nonpermanent form of inhibition. Ultimately, after several days of extinction training, spontaneous recovery is no longer observed; extinction is by then totally controlled by a permanent form of inhibition.

Drawing on the facts of extinction and spontaneous recovery developed in Pavlov's research, Hull built into his theory both forms of inhibition and, as a special bonus, showed how ${}_SI_R$ developed through the mediation of $I_R$. In particular, Hull endowed $I_R$ with the following set of properties: (1) Whenever a response is made, a certain amount of reactive inhibition is generated; (2) the greater the effort required to make the response, the greater the amount of $I_R$; (3) over massed repetitions of the response, $I_R$ accumulates; (4) $I_R$ is aversive, acting like a drive; (5) $I_R$ dissipates rapidly over time; and (6) the presence of $I_R$ inhibits repetition of the response that gave rise to it.

Properties 4 and 5 in combination underlie the development of ${}_SI_R$. That is, the dissipation of $I_R$ is analogous to the abrupt reduction of a drive in that it can serve as a reinforcer. What gets reinforced, and hence learned, is *not responding* (not performing R) in the presence of the stimulus, S, since it is not making R that permits $I_R$ to dissipate. So, even though $I_R$ itself is short-lived, it leaves a permanent residue in the form of conditioned inhibition.

*Reactive Inhibition*

By now it should be apparent how the concept of reactive inhibition could be enlisted in the explanation of spontaneous alternation. In the T maze, or Y maze, the animal has two competing response tendencies at the choice-point: turn left or turn right—i.e., make the turning response that leads the animal into one goal arm or the other. Having, for whatever reason, made one of those responses, say turning left, the animal faces the choice-point on the second trial with a certain amount of left-turning reactive inhibition and no (or virtually no) right-turning inhibition. In the absence of any other compelling basis for choice, the animal makes a right-turning response and hence alternates. SAB, in short, reveals the operation of reaction inhibition. Given the high esteem in which Hullian theory was held in the 1940s, and given its utter simplicity, it is no wonder that a reactive inhibition-based explanation of SAB was so appealing.

Note the fundamental differences between the Hullian acccount and its less formally stated predecessor. (1) For Dennis and Tolman, alternation reflected an approach tendency: in the service of fully exploring the maze, the animal was attracted to alleys it had not previously visited. By contrast, from the $I_R$ perspective the animal makes the alternate response on $t_2$ almost by default: it avoids repeating the prior response and takes the only available option.

Incidentally, one might ask why respond at all, either on $t_1$ or $t_2$? But that question simply was not raised in this context. The only question was which response (left or right turning) would be made. Of course, as any animal researcher knows, rats sometimes do stubbornly refuse to pick a goal arm, whether on $t_1$ or $t_2$. When that happens, we either wait the animal out, or treat that event as a "missing data point," to be filled in at a later date. It seems to me that a complete account of SAB needs a mechanism for getting the animal moving up to the choice point and into a goal arm as well as for determining which of two goal arms will be entered.

(2) For Dennis and Tolman, what was important was *where* the animal was in the maze; for the Hullian account what mattered was *how* it got there. The first approach can be identified as stimulus-based, the other as response-based.

The applicability of Hull's reactive inhibition concept to the explanation of SAB was formally articulated by Richard Solomon (1948) in an article on work and behavior based on his doctoral dissertation. Solomon actually offered a modified version of Hull's concept, which he believed to be an improvement. In essence, Solomon pointed to kinesthetic feedback from motor responses as the locus of the inhibitory process, rather than the musculature itself. In that way, reactive inhibition becomes less like fatigue and more like sensory adaptation; nevertheless, the inhibition originates in the animal's own motor activity and is not externally oriented. Thus, though Solomon's version treats reactive inhibition as "sensory," it still differs importantly from the Tolman-Dennis conception, which has animals adapting to environmental stimuli, not to feedback from their own responses.

It is easy to see how fatigued muscles may be harder to activate than fresh ones and how this difference can lead to SAB. Why differential adaptation of kinesthetic fibers leads to SAB is not nearly so apparent unless one recalls the assumption that reactive inhibition is endowed with drive-like properties and cumulates over massed repetitions. While neither Hull nor Solomon makes this point explicitly, the animal must be assumed to avoid or at least spread out repetitions of a given response in order to minimize the level of reactive inhibition, just as a rat might avoid entering a part of the maze associated with painful electric shock.

### *Intertrial Interval*

One obvious prediction from the reactive inhibition account, whether Hull's or Solomon's, is that the probability of SAB should decline as the interval between $t_1$ and $t_2$ increases, reflecting the dissipation of reactive inhibition over time. The data, in a rough way, supported that prediction. For example, in an early test of this hypothesis Heathers (1940) varied ITI between 15 and 120 seconds and found the expected inverse relationship. Other early studies (e.g., Dennis, 1939) also found the SAB rate to decline to chance with a one-to two-minute ITI, as did some experiments a decade later (e.g., Montgomery, 1951; Riley & Shapiro, 1952). However, markedly departing from the usual finding, Edward Walker (1956) reported significant SAB (about 0.75) for intervals up to 60 minutes and then a sharp drop to a chance level at 90 minutes. Possible methodological reasons for this dramatic discrepancy are suggested by Dember & Fowler (1958) and need not be repeated here. For present purposes the point is that SAB rate does decline with increasing ITI, as expected from the reactive inhibition account. On the other hand, for reactive inhibition to function, as it is presumed to in Hullian theory, in the development of conditioned inhibition, its dissipation should be very rapid. While Hull does not attach a specific numerical value to "rapid," results like Walker's seem embarrassing to a strictly interpreted reaction inhibition explanation of SAB.

*Effort*

Another property attributed to reactive inhibition is its relation to the amount of work or effort expended in making the response: the greater the effort, the higher the value of reactive inhibition, hence the higher the rate of SAB. While in principle readily subject to empirical test, this property of reactive inhibition was not adequately exploited, perhaps because of the difficulty of operationalizing the notion of work or effort in this context. Those few published studies pertinent to the effects of effort yielded results that, overall, were only weakly supportive of the Hullian account (e.g., Jackson, 1941; Mowrer & Jones, 1943).

In 1945–46 Solomon himself attempted to manipulate the effort variable by strapping weights onto the backs of his rats and using inclined maze arms. He reports that increased work, so operationalized, was reflected in the expected increase in rate of SAB, but the difference from the rate exhibited by suitable control animals was not statistically significant. Unfortunately, Solomon did not pursue that line of research any further.

To set the record straight, Dember & Fowler (1958), in alluding to those results, attributed them to Solomon's 1948 article, "The Influence of Work on Behavior." Actually, the data were never published; "they were reported only in conversations, because the results were not decisive" (Solomon, personal communication, 1988).

The paucity of supporting data on the work variable was, in the 1940s, especially unfortunate for the reactive inhibition account since, as Solomon pointed out, the relation between SAB and ITI, though predicted from reactive inhibition, was also consistent with the various explanations provided by Dennis (1939), Heathers (1940), and Tolman (1925).

"Actually, the demonstration of the relationship between effortfulness of task and avoidance of repetition of responses would be very significant. The effects of *time interval variations* were predicted fairly well by [other] conceptualizations. . . . In this case, Hull's theories add little. But the role of the *effort variable* is not a salient feature of the other theories. . . . Therefore, the effort or work variable becomes crucial in testing the alternative hypotheses." (Solomon, 1948, pp. 25–26)

Additional experiments reported in the 1950s did not change the picture with regard to the effort variable. Either effort variation showed little if any effect (e.g., Riley & Shapiro, 1952; Walker, Dember, Earl, Fawl, & Karoly, 1955) or employed a manipulation—massed, forced trials to one goal arm followed by a free trial—that by then could easily be interpreted within alternative frameworks (e.g., Estes & Schoeffler, 1955; Zeaman & Angell, 1953; Zeaman & House, 1951).

The most severe blow to a reactive inhibition account of SAB came from experiments in the 1950s that directly pitted that explanation against one based on the notion that animals were alternating stimuli, not responses.

## *SAB in the 1950s: The Decline and Fall of Reactive Inhibition*

Appealing in its simplicity, the reactive inhibition explanation of SAB was nevertheless found wanting. Effortfulness of responding did not prove a fruitful variable and the predicted effect of ITI, though supported empirically, was not specific to reactive inhibition. Any explanation that relied on some sort of memory trace to link $t_1$ and $t_2$ would also predict a decline in SAB with increased time between trials.

### *Cues for SAB*

The decade of the 1950s saw several devlopments in our understanding of SAB; chief among them was the demonstration of the primacy of external cues over responses or response-produced cues in the control of SAB. Along with this renewed emphasis on a stimulus-based account of SAB came a resurrection of exploration as the putative underlying motivation for SAB. At the same time, other explanatory concepts were suggested, and other questions were raised, such as the phylogenetic generality of SAB. All of these issues will be addressed in greater detail in subsequent chapters.

Two researchers, Kay Montgomery and Murray Glanzer, made the initial, major contributions to the demise of reactive inhibition as the explanatory concept for SAB. Their work was based on the same logic, which will be covered here in broad strokes and then elaborated in Chapter 2.

Consider the standard T maze (see Figure 1.1), with a starting stem leading to a "choice-point" at the juncture of the starting stem and the two arms of the T (usually referred to as "goal arms"). On $t_1$, the animal approaches the choice point and then, depending on the observer's theoretical orientation, "enters one of the goal arms," say the arm on the right or the white arm, or "makes a right-turning response," which carries the animal into the right-hand (white) goal arm. Incidentally, one might also observe the animal going "east" if its movement in the starting stem is in a northerly direction. For this animal to show alternation

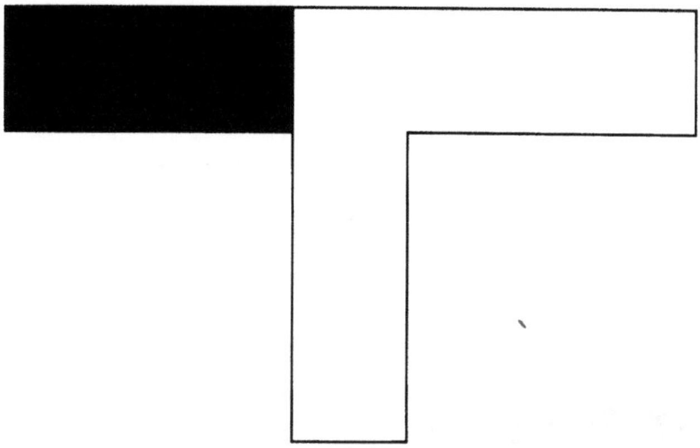

FIGURE 1.1. Plan view of T maze with a white and a black goal arm.

on $t_2$, it must enter the left (black) goal arm or make a "left-turning response." Entertain the hypothesis that either of these scenarios might be the correct way to characterize what is happening. The observed behavior pattern will be the same whichever is the proper account–i.e., response alternation or stimulus alternation. So, that way of investigating SAB does not enable us to distinguish between the two accounts, just as one might expect since both were designed to account for SAB as conventionally studied.

Montgomery and Glanzer found an elegant way to disentangle responses and stimuli, as illustrated in Figure 1.2. The standard T maze is transformed into a + maze by the addition of a second starting stem. On a given trial one of the two stems is closed off at the choice point, restoring the maze to a T configuration. Suppose an animal is given trial 1 using the starting stem so labeled in the figure. Let it turn right (enter the white goal arm). On trial 2 the starting stem so labeled is used, with the other blocked off. In essence, the animal approaches the choice point from opposite directions on $t_1$ and $t_2$. As a result, the response and external stimulus bases of SAB become unconfounded. Our rat that turned right and entered the white goal arm on $t_1$ is now faced with a choice between alternating responses (turning left) and repeating stimuli (reentering the white arm) or alternating stimuli (entering the black arm) while repeating responses (turning right again).

Data obtained by Montgomery (1952), Glanzer (1953b), and others (e.g., Walker, Dember, Earl, & Karoly, 1955; Dember & Roberts, 1958), unequivocally reveal rats to be stimulus alternators under the conditions just described. That outcome of the + maze studies, more than any other set of results, proved the undoing of reactive inhibition as a viable explanation of SAB in rats. The utility of that concept, or of analogous response-based notions, may, however, remain for alternation behavior observed in some lower species, especially invertebrates, as will be discussed in Chapter 3.

Just as the + maze experiments restored prominence to stimulus accounts of SAB, so too did they ultimately lend credence to the notion that SAB is a

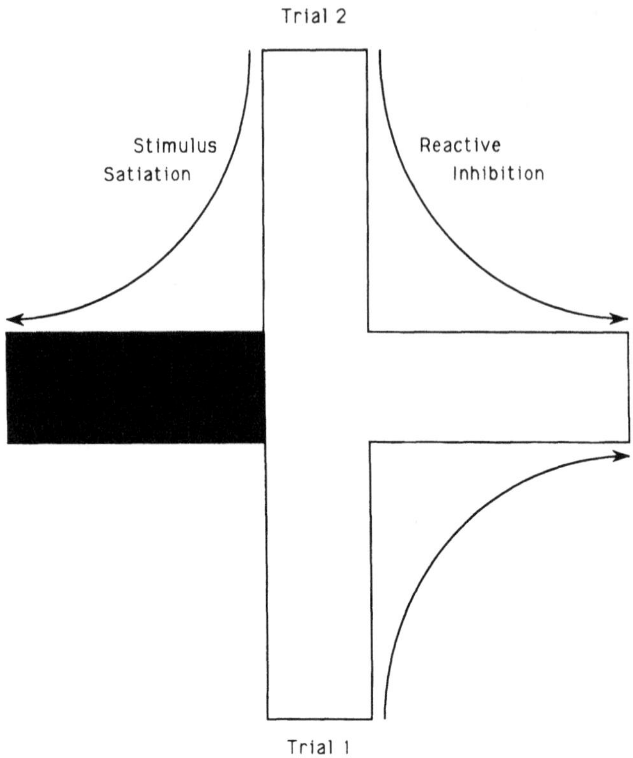


FIGURE 1.2. Plan view of a t maze. If, starting from the south on trial 1, the animal turns right into the white goal arm, then on trial 2, starting from the north, it is faced with a choice between turning left (going again into the white arm), thereby alternating responses (as predicted by the reactive inhibition account of SAB) or again turning right (into the black arm), thereby alternating stimuli (as predicted by the stimulus satiation account).

manifestation of an exploratory motive. The story of how that came about will be saved for Chapter 2. My own entree into this fascinating business occurred in the early 1950s, working with fellow graduate student Robert Earl in Edward Walker's laboratory at the University of Michigan. It was, in fact, through SAB that we were led to our ideas, formalized as the Theory of Choice (Dember & Earl, 1957), about exploratory behavior, attention, and preference. It will suffice here to note that we saw SAB as a special case of a more general response to change or novelty and the latter as closely related to the concept of stimulus and individual complexity. While those notions were percolating, we were also heavily involved, under Walker's direction, in a series of experiments designed to identify the exact locus of the stimuli to which the animals were alternating in T and + mazes. Again, saving details for the next chapter, our research pointed to stimuli within the maze as the predominant "cue" for SAB, with extra-maze stimuli next in salience, followed weakly by response-produced cues (Walker, Dember, Earl, & Karoly, 1955). To anticipate a bit, the general strategy employed was an extension of the logic of Montgomery and Glanzer, that is, to segregate otherwise confounded cues. Recall that starting the animal from opposite ends of the t maze on trials one and two enabled a separation of stimulus and response. But note that two gross types of stimuli are still entangled: those within the maze and those external to the maze (suggested earlier by our reference to the white (east) and black (west) goal arms. How those two categories of stimuli were rendered independent is another part of the story to be picked up in Chapter 2.

To summarize thus far, the early and mid-1950s saw the decline and fall of reactive inhibition as a satisfactory explanatory concept for SAB and some articulation of the locus of the stimuli (the "cues") to which the animal was alternating. What other developments in the decade of the 1950s are noteworthy?

*The Action Decrement*

As interest in SAB grew, the scope of the research also expanded, as did the variety of explanatory systems that were enunciated. One interesting theoretical development in this era was the articulation by Walker (1956; 1958) of his concept of the "action decrement" (initially called "reaction decrement"). In essence, Walker postulated a central inhibitory event, the action decrement, symbolized $I_C$ to contrast it with Hull's $I_R$ and an analogous term, stimulus satiation, symbolized $_SI$, proposed by Glanzer. Walker's aim was to transfer the hypothetical inhibitory process responsible for SAB from the periphery to a more central locus. Moreover, and more importantly, Walker wanted to tie SAB to those motivational processes, drive and reinforcement, that played such a critical role in the learning theories of that time. Finally, his model was designed to allow for interactions between motivational and temporal variables. In Chapter 2, we will closely examine some specific predictions generated by this model and the data offered in their support.

*Phylogenetic Comparisons*

With regard to purely empirical contributions, a trend was started early in the decade that was to flourish in subsequent years—the demonstration of SAB in animals other than rats. Except for some experiments on human subjects that revealed a tendency not to repeat a response on various guessing tasks or in situations involving difficult discriminations (e.g., Irwin & Preston, 1937; Luh, 1931; Robinson, 1934; Skinner, 1942; Telford, 1931; Thorndike, 1927; Wingfield, 1943), all of the research on SAB since Hunter's original observations had been done with rats as subjects. And the human experiments, though tantalizing in their implications, were conducted in ways that did not closely resemble the SAB paradigm. So a critical question remained: Is SAB restricted to rats (and hence of limited interest), or can it be observed in other species? One experiment (Lepley & Rice, 1952) did indeed find what looked like SAB in a single-celled organism, *Paramecium aurelia*. Subsequently, it has become abundantly clear that SAB is not confined to rats. Indeed, it, or something very much like it, appears throughout the animal kingdom. Not every species, of course, has been tested, but there is evidence for phylognetic generality. Whether a common explanation is applicable across species is another question, one that will be carefully addressed by Robert N. Hughes in Chapter 3.

*What is Chance?*

In their review of the SAB literature, Dember and Fowler (1958) raised the methodological issue alluded to earlier in this chapter, that is, what constitutes a chance level of alternation? Their concern was with how to deal with an initial preference or bias in the animals' $t_1$ choices. Absent such bias, then the chance level of alternation, $P_O$, against which to assess the reliability of any observed level, is 0.50. Dember and Fowler argued that if there is a strong bias, then $P_O = 0.50$ overestimates the expected level of alternation and hence underestimates the significance of the obtained level. Should there be a strong preference for one of the alternatives, say goal arm A or turning response A, then a "more reasonable estimate of chance alternation would take into account the distribution of Trial 1 behaviors. Thus, if 80% of the behaviors on Trial 1 were of Type A and 20% of Type B, the a priori probability of alternation, $P_O$, would be $1 - (0.80^2 + 0.20^2)$ or 0.32" (Dember & Fowler, 1958, p. 413). Such a formula was used by Sutherland (1957) in analyzing his SAB data. Subsequently, Koppenaal (1962) argued that the proposed correction for bias was seriously flawed in that it assumed that biases were homogeneous in a group of animals; i.e., that the group average provided a proper estimate of each subject's initial bias. Moreover, the Dember and Fowler formula also failed to acknowledge that the empirical probability was unlikely to be a perfectly reliable estimate of the underlying hypothetical probability. In either case, using the proposed formula would "result in systematic underestimation of the true $P_O$" (Koppenaal, 1962, p. 666). Koppenaal was unable to recommend a theoretical solution to this problem, but see Manning (1973a; 1973b) for a potential procedure rooted in experimental design. And Douglas and Isaacson (1965) provided convincing empirical evidence that a satisfactory bias-estimate for individual subjects could, indeed, be derived from the group data. Finally, returning to Dember and Fowler, a preferable alternative to correcting statistically for bias "would involve careful control of the experimental conditions so as to minimize unequal distributions of Trial 1 behaviors" (Dember & Fowler, 1958, p. 413).

While there is no way to attribute causality here in the absence of a proper control condition, the publication of the Dember and Fowler review was followed, in the decade of the 1960s, by a burgeoning of research on SAB and a further broadening of the issues investigated. The major contribution of that era was undoubtedly made by Robert Douglas (1964, 1966a) in his doctoral dissertation and then in follow-up studies.

## *SAB in the 1960s: The Plot Thickens*

### *Identifying Specific Cues*

Working under the direction of Robert Isaacson at the University of Michigan, Douglas set out in his dissertation to identify the specific stimuli with respect to which rats alternated. He wanted to analyze into their constituents the intra- and extra-maze cues grossly specified by Walker and his collaborators. What exactly were those intra-maze cues and those extra-maze cues? Were there, indeed, multiple cues, or perhaps just one or two in each set? The answer to those questions, and the clever procedures that Douglas devised to come by that answer, will be detailed in Chapter 2. Here, let it simply be asserted that Douglas concluded that there were two and only two alternation cues, one within the maze and one beyond the maze, neither of which had anything to do with vision. Labeling goal arms as black or white, for example, may have been convenient for experimenters, but the stimulus attributes to which those labels referred meant little to the rats. Perhaps that should not have come as such a great surprise since rats, though capable of visual brightness and pattern discriminations, are notoriously shortsighted. But not nearly so much as those in the 1950s who simply took it for granted that what was for us such a salient feature of the maze was also so for our subjects.

### *Brain Structures*

At the risk of seeming immodest, I am constrained to note what was probably the second most significant finding of that period—the demonstration by Roberts, Dember, and Brodwick (1962) of the role of the hippocampus in the control of both SAB and maze exploration. That experiment, which pioneered what was to become an extensive literature on brain mechanisms in SAB and related phenomena, was conceived within a framework that required some sort of memory trace linking $t_1$ and $t_2$. The notion of a mediating memory trace, of course, was central to much of the early research on intertrial interval. In the present case, the hypothesis to be tested was that serious damage to brain centers critical for the functioning of memory processes, especially those implicated in relatively short-term memory, would be revealed in a reduced level of SAB. Though virtually no research had been done at the time on the role of the rat hippocampus in memory functioning, there was strongly suggestive evidence in the human clinical literature (e.g., Penfield & Milner, 1958). So, Roberts, Dember, and Brodwick proceeded to prepare rats with large bilateral hippocampal lesions and test them for both SAB and exploratory behavior. Control subjects were also run, either sham operates or animals with neocortical lesions. As predicted, the hippocampally lesioned rats showed a marked reduction in SAB, on average to a chance level, whereas the two control groups revealed significant levels of SAB.

The data with regard to exploration were also consistent with expectation. All animals were allowed 10 minutes of free exploration in the T maze. All groups started at the same level of activity. However, whereas the control animals showed the usual habituation of their exploratory activity over the 10-minute testing period, the level of activity exhibited by the hippocampal rats remained high. It is as though for the latter animals, lacking adequate short-term memory, the maze retains its novelty despite recent repeated exposures to its elements.

Since the study just described was reported, alternative accounts of the deficit suffered by rats with hippocampal lesions have been offered, though the basic empirical results have withstood replication attempts. A thorough discussion of these issues is provided by Robert Douglas in Chapter 5.

*Neurochemistry*

Work on specific neurochemical factors in SAB also saw its beginnings in the 1960s. As far as I can tell, the initial study of this sort was reported by Douglas and Isaacson (1966), who administered scopolamine, an anticholinergic agent, to rats and noted a substantial decline in the rate of SAB, to a level not significantly above chance. The argument was that scopolamine acted analogously to hippocampal lesions, which, as noted earlier, severely disrupt the alternation tendency. Since that line of investigation began about two decades ago, a great deal of additional empirical information has been attained and a better understanding achieved of the complex interactions through which the drug-behavior relations are realized. One of the major contributors to this literature, Larry Kokkinidis, offers a thorough presentation of this part of the SAB story in Chapter 6, which also covers the more general issue of the neurochemistry of habituation. It is my impression that it was primarily drug and other neurobiological studies that helped maintain interest in SAB during the 1970s and 80s, a period during which controversy, and hence excitement vis à vis more purely psychological issues (e.g., cues and motives), began to wane.

*Ontogeny*

One other significant topic came to the fore in the 1960s, that is, the ontogeny of SAB. The typical subjects in SAB studies are mature rats. One might reasonably ask whether the appearance of SAB follows a developmental course or whether it is full blown as soon as the animal can get about. Kirkby (1967) provided data to show a lack of SAB in locomoting infant rats. Subsequent experiments identified the age—about 30 days—at which adult levels of SAB are manifested under standard testing conditions. Moreover, the age of onset of SAB closely corresponds to the age when the dentate gyrus of the rat hippocampus becomes functional (Altman & Das, 1965), suggesting a causal relationship. But to add further intrigue to the story, a decade later Smith and Spear (1978) reported significant levels of SAB in 16-day-old rats if shavings from the home cage are present in the maze. Norman Spear and James Miller provide an extensive review of the ontogeny of SAB in Chapter 7.

*SAB as Indicator Response*

When a behavioral phenomenon is well understood and there is a sufficiently large corpus of supporting empirical data, it often becomes tempting to use that phenomenon as an indicator, or tool, for studying other issues. Thus it has been with classical and operant conditioning. Some tentative moves in this direction were made in the 1960s vis à vis SAB. For example, if normal SAB is dependent on an intact short-term memory, then SAB might prove useful in investigating factors that are hypothesized to impair short-term memory. Similarly, if SAB is sensitive to the similarity between the alternate goal arms, as postulated in Glanzer's (1953a) model, then SAB might be employed as an indicator response for investigating similarity. As a participant in a symposium organized by Glanzer for the 1960 meetings of the American Psychological Association, I (Dember, 1960) suggested both possible uses of SAB, referring specifically to a study by Dember and Millbrook (1956) and to the data on effects of hippocampal lesions that were subsequently reported by Roberts et al. (1962). This way of approaching SAB, that is, as a tool, received some attention in subsequent decades and is elaborated in Chapter 8 by Robert Douglas.

In summary, the decade of the 1960s was a prolific one for SAB. Issues raised earlier were vigorously pursued, e.g., the identity of the cues for alternation. New lines of investigation were opened up, in particular the comparative and developmental study of SAB and the brain structures and neurochemical factors implicated in the expression or inhibition of the alternation tendency. History did not stop in 1969, but the stage was by then set for work that is better treated as contemporary and organized not around era, but around specific themes, as will be reflected in the following chapters.

# 2
# The Search for Cues and Motives

William N. Dember
UNIVERSITY OF CINCINNATI

Spontaneous alternation behavior (SAB), as exhibited by mature, intact, tame animals, is among psychology's most reliable phenomena. It also appears to be incredibly simple, and yet it has posed problems for some of our most inventive and insightful researchers. Why do animals alternate? Exactly what is it to which they are alternating? Those two questions, intimately intertwined, come up again and again. A direct answer to the latter (the "cues" question) is a lot easier to achieve in principle than one to the former (the "motive" question). Cues can be uncovered through clever, though straightforward analytic procedures; identifying motives is much more dependent on global inferences and an intuitive feeling for the "rightness" of a hypothesis. In this chapter, I will address both questions, beginning with the easier one. Recognize, however, that the two questions are not independent and that answers to one can have profound implications for the other.

## *What are the Cues?*

### *Where are the Cues?*

Recall that both Kay Montgomery and Murray Glanzer showed rats to be stimulus alternators, not response alternators. Hull's reactive inhibition concept, so appealing at first, simply could not accommodate the results of their studies. And so the question shifted from "Is it reactive inhibition or something else that lies behind SAB?" to "What is that something else?" That is, granted that rats are alternating with respect to stimuli, not responses, what are those stimuli or "cues" or might it be that any discriminable stimuli will suffice? In his own attempt at developing a formal model that could serve as a replacement to the reactive inhibition account of SAB, Glanzer postulated a construct, stimulus satiation, that applied to any stimuli to which the animal was sensitive: "Each moment an organism perceives a stimulus-object or stimulus-objects, A, there develops a quantity of stimulus satiation to A" (Glanzer, 1953a, p. 259). Glanzer's model will be taken up in greater detail later in this chapter, when the question of motives is addressed. It is pertinent here because it did suggest some experimental manipulations that bear directly on the issue of cues.

The most immediate reaction to the identification of "stimuli" as the focus of SAB was a set of experiments designed by Edward Walker and his students intended to tease out the specific loci of those stimuli to which the rats were alternating. Were those stimuli inside the maze or external to the maze, or perhaps even to some extent associated with feedback from the animal's turning response, as Richard Solomon (1948) had proposed?

Montgomery and Glanzer both employed a + maze in their experiments, enabling them to segregate stimulus from response sources of SAB. Walker et al. employed a rotatable + maze, which allowed the separation of intra- from extra-maze stimuli. The logic is as follows:

1. First, consider the case when the + maze is used as though it were the standard T (see Figure 2.1, panel A). Animals are given two trials from the same starting alley and face the same alternatives at the choice point. Under those testing conditions, internal cues, extra-maze cues, and response-produced cues are completely confounded. The animal alternates, but it is not possible to tell with respect to what.
2. The + maze is used as Montgomery and Glanzer employed it (Panel B). That is, animals are introduced into one starting stem on $t_1$ and into the other stem on $t_2$. Under those circumstances, one can segregate response-produced from intra- and extra-maze cues, but the latter two sets are still confounded. Thus, on $t_1$ let a rat be introduced into the south starting stem and enter the white (right, east) goal arm; on $t_2$ it is started in the north stem; if it alternates, it does so by entering the black (right, west) goal arm. That is, it alternates with respect to goal arm brightness (black or white) and/or with respect to the part of the room (building, world) that it approaches, while at the same time repeating its prior turning response (right).
3. The + maze is rotated through 180° between $t_1$ and $t_2$ (Panel C). Again, let it enter the south starting stem on $t_1$ and enter the white (right, east) goal arm. For $t_2$, the maze is rotated as indicated and the animal starts from what is now the north starting stem. Note that to alternate with respect to intra-maze stimuli (represented here as white and black, though the relevant intra-maze cue may in fact not be visual), the rat must enter the black arm. But to do so, it also alternates with respect to its prior turning response (right on $t_1$, left on $t_2$), and it must repeat (move into the location of, or in the direction of) the extra-maze cues designated as east.

Using such logic and operations, Walker, Dember, Earl, & Karoly (1955) concluded that the potency of the three classes of cues was ordered as follows: intra-maze, extra-maze, response-produced. In a follow-up experiment, a modified + maze was constructed that placed a banked turn at the choice point, with the intent of making left- and right-turn response-produced cues more distinguishable and more salient (Walker, Dember, Earl, Fawl, & Karoly, 1955). That manipulation did succeed in increasing the amount of SAB attributable to response-produced cues, largely at the expense of extra-maze cues.

Taken together, the results of the two experiments just described indicate that (1) there are two important classes of cues for SAB in the standard apparatus,

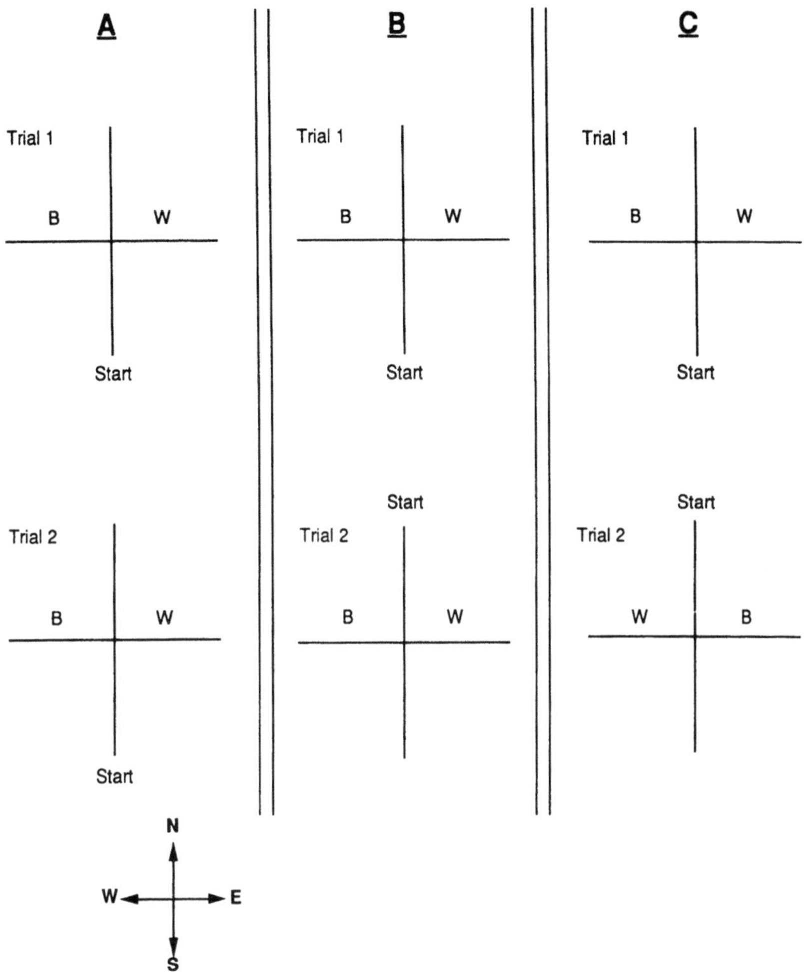


FIGURE 2.1. Various maze arrangements for studying SAB. Panel A shows the typical paradigm, where the apparatus is used as a T maze. In Panels B and C, the apparatus is used as a t maze. The configuration in Panel C differs from that in Panel B in that the maze is rotated 180° between trials 1 and 2, enabling the separation of intra- from extra-maze stimuli.

intra- and extra-maze; (2) response-produced cues are typically negligible, but if increased in salience can be rendered modestly effective; and (3) intra-maze cues are more important than extra-maze cues. The third point was subsequently to be called into question by Douglas (1966a), as we shall see, though it certainly did accurately characterize the Walker et al. data.

### *Does Passive Exposure Work?*

Before turning to Douglas's experiments, I want to note another line of research that followed from Glanzer's model, again conducted in Walker's laboratory.

Glanzer's satiation model assumes that whenever an animal perceives a stimulus, a certain amount of satiation develops, vis à vis that stimulus, and such satiation reduces the animal's responsiveness to that stimulus. Glanzer further postulates that satiation dissipates over time (just like reactive inhibition in Hull's theory).

It seemed implicit in Glanzer's formulation, if not explicit, that the stimulus exposure did not have to result from the animal's bringing itself into contact with that stimulus by traversing the starting stem of a maze and then entering the goal arm containing that stimulus. Satiation ought to be able to occur anywhere, with the animal's behavior in the maze serving to make manifest the presence of that satiation. Specifically, suppose a rat is placed for a few minutes in a white box separate from a T maze with a white and a black goal arm. After being thus exposed to whiteness, the animal is introduced into the maze. Having been satiated for whiteness by being exposed to it outside the maze, the animal should, on its first trial in the maze, enter the black arm.

A straightforward test, based on the preceding argument, was conducted by Walker, Dember, Earl, Fliege, and Karoly (1955). Their data failed to confirm the satiation-based prediction. Additional unpublished studies in Walker's lab also yielded results inconsistent with expectations from Glanzer's model. But the notion of exposure-generated satiation was too appealing to drop. Perhaps it was being tested in too simplistic a manner. Robert Earl suggested that to be effective such exposure might need to take place in the context of the choice point, where all cues relevant to the animal's ultimate choice are concentrated. Pursuing that logic, Kivy et al. (1956) designed an experiment in which rats were allowed to move up the starting stem to the choice point. There they were confronted by Plexiglas doors, which allowed inspection of the goal arms, but prevented the animals from entering either arm. Each rat was given 15 minutes in the maze, not all of which, of course, was spent in the choice-point region. Then, the rat was gently removed from the maze, the Plexiglas doors were removed, one goal arm changed from black to white or white to black, and the animal was returned to the starting stem for a "choice" trial. The results revealed a significant preference for the "nonsatiated" goal arm. That is, if on the exposure trial both arms were black, the animals tended to enter the white arm on the choice trial, and vice versa.

Those results seemed to offer vindication of Glanzer's satiation-through-exposure postulate as slightly modified to incorporate the importance of the context in which the exposure was taking place. But uneasy with that conclusion, I hit, almost inadvertently, on a different interpretation of the results.

In describing the Kivy et al. experiment to a class, I found myself reversing the maze conditions that prevailed on the test and exposure trials. That is, on the exposure trial I had the goal arms different in brightness, one black, the other white, while on the test trial I had them the same, both black or both white. Noting my own error, I corrected it, and went on with my lecture. But the error perseverated in memory, and I began to wonder what would happen, and especially what it might mean, were the experiment to be conducted in accord with my erroneous description.

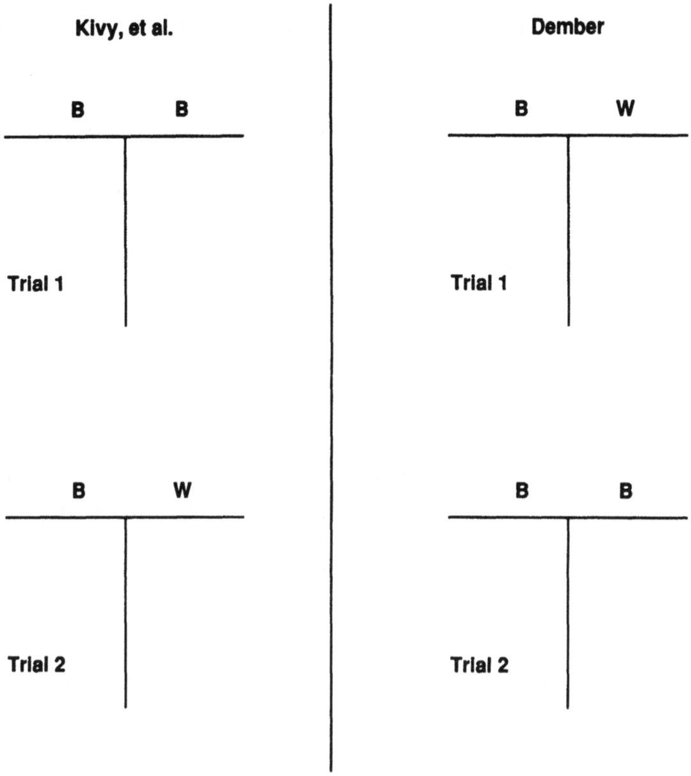


FIGURE 2.2. Examples of maze arrangements for the Kivy et al. (1956) choice-point exposure experiment and the Dember (1956) response-to-change experiment.

The two designs are shown in Figure 2.2. In the Kivy et al. design, the animal is faced on the choice, or test trial, with an arm of a brightness to which it had presumably been satiated during the exposure trial and an arm of a brightness maximally different from that to which it had been exposed, that is, a black arm and a white arm. The animal chooses (enters) the nonsatiated arm. But the animal could also be said to have chosen the arm that was changed in brightness between exposure and test. Perhaps the rat is responding to change, not to satiation. Now consider the other possible design. Here, the rat is exposed to a black arm and a white arm and then offered a choice between two white arms, for example. Satiation per se provides no basis for choice, since the animal is equally satiated for both (white) arms. However, one of those white arms had been black on the exposure trial and was changed to white for the test trial. The other arm remains as it had been. The hypothesis of a choice based on response to change would permit a clear prediction in this design, whereas the satiation hypothesis would be mute.

The experiment suggested by these ruminations was quickly run, and it was found that 17 of 20 animals entered the changed arm (Dember, 1956). On the basis of these results, which have been replicated by several other investigators (e.g., Fowler, 1958; O'Connell, 1964; Walk, 1960; Woods & Jennings, 1959), I argued that animals in the Kivy et al. study, and by extension in the standard SAB experiment, were not just avoiding a satiated arm and by default entering the non-satiated arm. Rather, they were being drawn to the changed, or novel arm. That way of thinking about SAB made it easy to see it as a specific instance of a general tendency to be attracted by, attend to, and explore novel or complex objects or events (see, for example, Dember & Earl, 1957; Dember, 1965; Richman et al., 1987). I will specifically address the response-to-change concept in the second part of this chapter, which is concerned directly with the issue of the motivational basis of SAB.

### *The Douglas Discovery*

In the previous chapter, I asserted that the major breakthrough of the 1960s was the discovery by Robert Douglas (1966a) of *the* two cues for SAB, noting only that one of them was within the maze and the other external to it. It is time to examine Douglas's research in greater detail and to lay out his grounds for specifying the nature of those two cues.

Prior to conducting the several experiments that comprised his dissertation, Douglas noted some reasons for skepticism about the role of visual cues in SAB. Aside from rats' notorious near-sightedness mentioned earlier, Douglas pointed to research by Dember (1958) and Dember and Roberts (1958) revealing normal levels of SAB in rats that had been rendered blind by removal of their eyes. Such results indicate, at the least, that visual cues are not essential for SAB. But if not visual cues, then what? To answer that question, Douglas extended the logic of Glanzer, Montgomery, and Walker, et al. by using various maze conditions that permitted the potency of single cues to be assessed. Of course, it order to tease out particular potential uses, one first needs some sense of what they might be. Empirical data by themselves do not substitute for thinking and intuition.

Douglas started with the three-fold classification of cues developed by Walker and his associates.

> "In the current experiment the three general classes of stimuli are defined as follows: Intra-maze cues are those with which the animal could conceivably come into direct contact. This includes the entire inner surface of the maze as well as the immediate surface of the floor (usually paper). Extra-maze stimuli are those having their origin outside the confines of the maze. This includes the sub-floor under the maze, and the visual, olfactory, and auditory stimuli of the environment of the maze. Response stimuli are those generated by the actual muscle and joint movements involved in making a turning response." (Douglas, 1964, p. 23)

One major innovation introduced by Douglas was the use of two mazes, two adjacent, but separate, testing rooms, and removable paper flooring on which the rats ran. A schematic diagram of the testing rooms, with the two mazes in their usual positions, is given in Figure 2.3. Both windows shown in the diagram were completely shaded by blackout curtains, and the wall opposite the choice point of

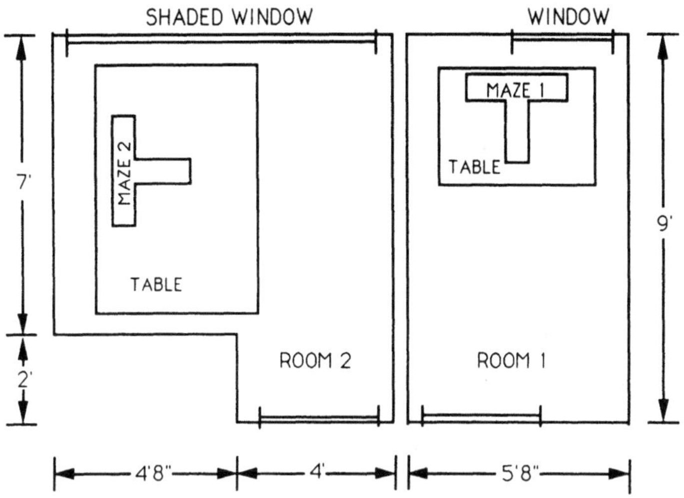


FIGURE 2.3. Schematic drawing of the testing rooms and maze configurations used by Douglas (1964) in his search for the cues used by rats in the SAB paradigm.

the mazes, as depicted on the figure, was covered by a continuous strip of opaque paper. Further specific details of procedure can be found in Douglas (1964) as well as a closely argued rationale for determining the proper value of the chance level of alternation in any given experiment.

First, some initial findings. (1) To assess the rate of SAB with all cues present, 48 male, hooded rats were run on two successive trials, with a brief ($<30$ second) ITI, in the same maze, in the same room, with the same paper flooring in place. This test was conducted on seven days, interspersed among days on which variations from the standard procedure were employed. On any given day, half of the S's were run in Room 1, half in Room 2. The overall rate of SAB was 80.4%, about the value typically reported. (2) To assess the role, if any, of response-produced cues, on four days animals were given $t_1$ in one of the two mazes and $t_2$ in the other. Since there were no intra- or extra-maze cues in common between $t_1$ and $t_2$, the only possible cues connecting $t_1$ and $t_2$, were "within the animal" in the form of response-produced stimuli. Consistent with previous research, the rate of SAB was 51.0%, not different from chance. A second experiment addressing this same point used a different maze configuration, one where the animal's first choice led it to a second choice point, with goal arms orthogonal in direction to the two initial goal arms, analogous to the multiple-Y configuration used in the early studies by Dennis and colleagues (see Chapter 1). One special feature of the multiple choice point procedure for studying response alternation is that it requires no handling between choices. These results also revealed no evidence of the operation of response-produced alternation cues: the rate of SAB was 49%. So, Douglas obtained results consistent with previous findings when all cues were present and when only response-produced cues were available.

What about cue specificity? The full story here would require more space than it can be given in this chapter, so I will restrict the remainder of the discussion to a few key experiments.

To assess the potency of intra-maze cues in isolation, Douglas gave his 48 rats $t_1$ in one room and then transported that maze, with its paper floor, to the other room for $t_2$. The alternation rate, based on two such tests per animal, or 96 tests in all, was 61.5%, significantly above chance, but also considerably less than that found with all cues present (approximately 80%). The latter comparison suggests the existence of other effective cues outside the maze.

In order to track down those potential extra-maze cues, two tests per rat were given in which $t_1$ and t2 were in the same room, with the $t_1$ and $t_2$ mazes located in the same place and oriented in the same manner. However, one maze was used on $t_1$ and the other maze on $t_2$, with new paper flooring inserted between trials. Hence, only extra-maze cues were available to guide the animals' choices. Under these circumstances, extra-maze cues only, the rate of SAB was 75%, significantly higher than both chance and the 61.5% rate obtained in the intra-maze-cues-only condition, but not significantly lower than the rate obtained with all cues present (80.4%). Douglas provides a convincing explanation for the disparity between these results (higher SAB to extra- than to intra-maze cues) and those reported by Walker and associates, who found intra-maze cues to be the more potent. In essence, he argues that the + maze, as used by Walker et al., underestimates the effectiveness of extra-maze cues. He also neatly disposes of another question that might arise: Why does the all-cues condition not yield a significantly higher rate of SAB than the extra-maze-cues-alone condition? Douglas shows that when "true" SAB rates are computed for the intra- and extra-maze cues conditions from the obtained data, which are inflated by "false positives," and those true values are added probabilistically, the expected rate of SAB for the all-cues condition is 81.75%, just about the value actually obtained. On the basis of that analysis, Douglas concluded that the intra- and extra-maze cues act independently and together account for all of the SAB observed under ordinary testing conditions. At this point, then, the issue is joined: What exactly are those cues?

To find out about specific intra-maze cues, Douglas assessed the impact of three potentially relevant sense modalities: vision, touch, and smell. Recall the previous researchers had tacitly assumed that "intra-maze" meant "visual." As noted in Chapter 1, Douglas ruled out vision as a likely source of intra-maze cues, again using the general strategy of manipulating testing conditions so as to isolate visual cues as the only available basis for linking $t_1$ and $t_2$ and measuring SAB rate under those circumstances. Specifically, he conducted experiments to assess three kinds of visual cues: (1) brightness differences (black versus white) between the linings of the walls of the two goal arms; (2) pattern differences between wall linings (circles versus stars); and (3) other possible visual cues that were specific to the maze. For the brightness and pattern experiments, the same general procedure was followed. Two mazes were set up in room 2, oriented at right angles to one another. Two tests were conducted on 40 rats, yielding 80 tests in all. On half the tests in the brightness study both mazes had the left goal arm lined with black cardboard and the right with white; on half, the left arms were both white and the right arms black. The same logic was used for the visual pattern (stars versus circles) experiments. In the latter case, pilot work revealed that rats are capable of learning to discriminate between the two patterns. Decades of research have demonstrated rats' ability to make a black-white discrimination. In addition, these experiments revealed a significant, approximately 2 to 1 preference on $t_1$ for the black arm, again indicating the animals' ability to discriminate between the black and white stimuli. There was also a modest right-turning preference. The results of both the brightness and pattern experiments (where there was no $t_1$ preference) showed no significant alternation tendency. Thus, even though the rats could discriminate between the visual characteristics of the two goal arms, they used neither brightness nor pattern as a basis for alternating—disconcerting not only for those who had assumed all along that that was what was happening, but especially for those like me, who had assumed that the motivation for alternating was so powerful that SAB would occur whenever circumstances permitted. More on that later.

To get at maze-specific cues, which might include visual as well as other sensory differences between goal arms (texture, odor, etc.), the same maze was used on both trials, but was moved from one room to the other between trials, and fresh paper flooring was provided between trials. Forty rats were given one test under these conditions, and the SAB rate was 47.9%, not different from chance. Thus, not only does it appear that intra-maze visual cues play no role in SAB, there also seems to be no role for any cues associated with the maze arms per se. But clearly there are effective intra-maze cues.

What about tactile cues? Douglas varied the texture of floor inserts. In sum, though the rats revealed their ability to discriminate between wire-mesh floor inserts and smooth boards by showing a strong preference for the mesh flooring, they did not show a significant SAB rate when differential flooring provided the only possible intra-maze cue.

Perhaps it is time to get directly to the point. What Douglas came to realize was that the key to the mystery lay (literally) on the paper flooring. With the same paper flooring in place on $t_1$ and $t_2$, the SAB rate was 65.6% when different mazes were used and the two trials were conducted in different rooms (with the two mazes oriented at right angles to one another, as depicted in Figure 2.1). That value, which was based on 96 tests (48 animals, each tested on two occasions), is significantly greater than chance and about the same as that reported earlier (61.5%) with all intra-maze cues present.

Since the tactile-cues experiment mentioned earlier failed to find evidence of SAB with respect to different floor textures (wire mesh versus smooth boards), despite the rats' ability to discriminate between the two floor textures, it is clear that the animals were not using difference in the paper flooring itself in the two goal arms as a basis for alternation. Indeed, the flooring was actually a large sheet of paper placed under the maze and it is extremely unlikely that there were discriminable differences between the two sections of the single sheet that happened to fall under the two maze arms.

The only sensible inference is that the rats were modifying the paper flooring in the $t_1$ goal arm as they walked on it, most likely by leaving on it some kind of odor trail. That was Douglas's inference, and there has appeared in the subsequent two and a half decades no evidence to indicate anything to the contrary. Thus, it is generally accepted that the major, if not the only, intra-maze cue for SAB is an odor trail laid down by the animal on the maze floor. It is still an open question as to whether additional intra-maze cues might be effective under some circumstances (see, for example, Eisenberger, Myers, Sanders, & Shanab, 1970; Hughes & Seanberg, 1970; Pate & DeLoache, 1970; Richman, Chappell, Crowe, Perry, & Redfearn, 1970), and of course what is true for rats (indeed, mature, male, hooded rats) might not hold for other animals.

Two additional studies comprised the set that was devoted to isolating the intra-maze cue(s). In one, Douglas placed gauze pads near the choice point in each arm of the T maze. One of these pads had been left in the rat's cage and presumably had been impregnated with the rat's own odor; the other was fresh. Using appropriate testing conditions, Douglas concluded that the rat's odor per se was implicated neither in the animal's $t_1$ choice nor in the rate of SAB. In a second study, which simply involved post-hoc data analysis, Douglas looked at rats' behavior when confronted with another rats' odor trail and concluded that odor-trail alternation occurs only with respect to a given rat's own odor trail, not to that laid down by another animal. There is a bit more to say about odor trail, especially with regard to this last point, but I will put that off until the rest of the Douglas saga is told. There remains the issue of identifying the extra-maze cue.

For people who work closely with rats, identifying odor trail as a pertinent factor in their behavior may not be all that surprising. Even tame rats will defecate or urinate in the maze, and they generally run with their bellies close to the floor. There is ample opportunity for them to lay some kind of personal scent on the bottom of the maze.

Specifying the extra-maze cue is a different story. What Douglas discovered is not the least bit obvious, even in hindsight. Indeed, having isolated the extra-maze cue, Douglas and others whose work followed had great difficulty in finding completely satisfying language with which to label it.

It should be apparent that I am alluding to a cue that does not fit into some familiar sensory rubric, such as visual, olfactory, or auditory. For convenience, call it direction of movement, or simply directionality. The procedure Douglas employed to uncover that cue followed the logic of his other experiments.

First, Douglas conducted two experiments to see whether SAB occurred with regard to visual extra-maze cues. For this purpose he constructed a large box that fit over the maze. On the ceiling of the box, clearly visible from within the maze, red and white striped patterns were so placed that they ran parallel to one goal arm and perpendicular to the other. These patterns were in place in both trials; otherwise there were no cues in common between $t_1$ and $t_2$ (except that in the second experiment in this set, the same maze was used on $t_1$ and $t_2$). That is, $t_1$ and $t_2$ were conducted in different rooms, with fresh paper flooring on $t_2$, and with the $t_1$ and $t_2$ mazes at right angles to one another. In neither experiment did the SAB rate differ from chance.

Following the experiments on visual extra-maze cues, Douglas assessed the effectiveness of what he called "deep floor" or "vibratory" cues (following Shepard, 1959), odor cues, auditory cues, and all of those cues combined. In no case did SAB reach a significant level, including the condition with all cues combined. So, what was the elusive extra-maze cue?

Douglas reports that, merely on a whim, he decided to reorient the two mazes. Whereas in the previously described experiments the two mazes had been oriented perpendicularly (see Figure 2.4, left panel), he now set them parallel (right panel). Using an auditory cue (a buzzer) outside the end of one goal arm (as in his previous auditory cue experiment), he tested rats in the parallel-maze configuration (right panel in Figure 2.4), still convinced that the auditory cue was not unimportant. And sure enough, the SAB rate came in at 77.1%, well above chance and about the SAB rate (75%) found for extra-maze cues in the very first test of the potency of extra-maze cues. But then Douglas repeated the experiment without the differentiating buzzer, and the SAB rate, though somewhat lower (70%), remained significantly above chance. Since $t_1$ and $t_2$ were conducted in separate rooms, Douglas dubbed this mysterious cue the "extra-room" cue.

There then followed a series of experiments to pin that cue down more firmly. In the initial experiment in this series, the hypothesis was tested that extra-room noise might be serving as a cue. Data were collected both on a weekday, when the outside noise level was high (e.g., from the elevator and the ventilation system) and on Sunday, when the lab building was quiet, with even the ventila-

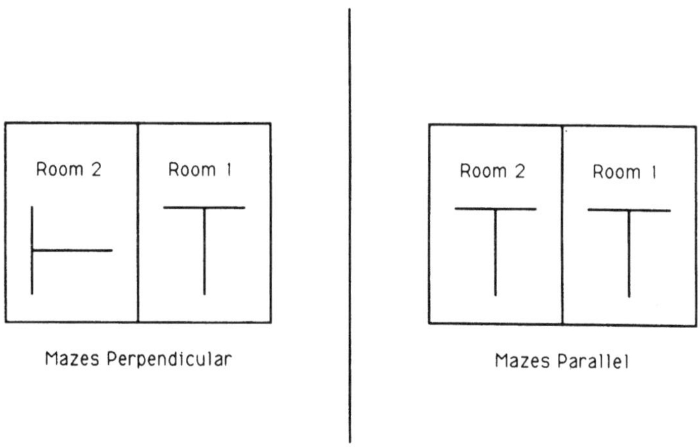


FIGURE 2.4. Perpendicular (left panel) versus parallel orientation (right panel) of mazes in Douglas's (1964) research. See text for explanation.

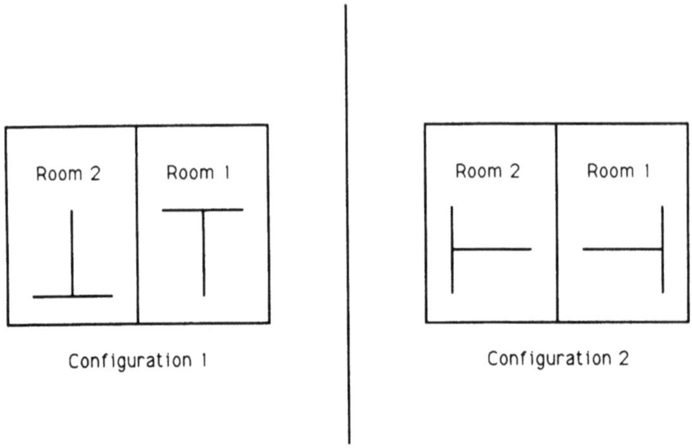


FIGURE 2.5. Further variations on maze orientation in Douglas's (1964) research. See text for explanation.

tion system shut down. There was no difference in SAB rate between those two days, with the Sunday results showing a slightly, though not significantly, higher (79.2%) rate then the weekday 77.6%). In another experiment it was shown that the extra-room cue remained potent (SAB rate = 81.3%) when the two parallel mazes were rotated 90° from their orientation in the experiment depicted in Figure 2.2.

Before proceeding further, Douglas decided to check whether response alternation, ruled out earlier, might have surfaced in the animals' reportory, which, if it had, could account for what looked like SAB to the elusive extra-room cue. But response alternation remained at a chance level.

Thus far, the extra-room cue had revealed itself in experiments in which the two T mazes were parallel *and* oriented in the same direction. What if they were still parallel, but oriented in opposite directions, as depicted in Figure 2.5? In these instances alternation to the extra-room cue would require response repetition, whereas in the earlier experiments alternation to the extra-room cue was compatible with any tendency, as weak as it might be, to alternate responses.

With the two mazes parallel and located in different rooms, but pointing in opposite directions, the SAB rate was 61.5%, lower than when the mazes pointed in the same direction, but nevertheless significantly greater than chance. In a subsequent experiment, with the two mazes located in the same room and parallel to one another but placed back to back (in a + configuration), SAB rate was 63.5%, again significantly above chance, but lower than when the mazes were parallel, in different rooms, and pointing in the same direction.

Douglas argued that rats, given the opportunity, try to alternate the direction in which they move, using vestibular feedback to help orient themselves in space. However, this effort can be disrupted if, through differences in $t_1$ and $t_2$ maze orientation, the rats are confused about the direction in which they are moving on $t_2$ (or in which they had traveled on $t_1$). Hence, the significant, but reduced, rate of extra-room cue SAB when the two mazes are parallel but pointing in opposite directions. To test out this interpretation Douglas employed manipulations designed specifically to produce the orientation disruption that he inferred was responsible for the reduced rate of SAB in the study.

In one experiment, Douglas rotated the animals 360° while on their way from room 1 to room 2. With different but parallel mazes in the two rooms, the SAB rate was 64.6%, not quite significantly lower ($p < 0.07$) than the rate obtained in prior experiments without the rotation manipulation. Using a more extreme procedure, giving the animals between trials "8 vigorous turns (of less than 360°) in alternating directions" (1964, p. 59), he obtained a rate of SAB of 53.1%.

Now the question arises as to whether the vigorous turns are emotionally, rather than spatially, disruptive, a pertinent consideration in light of the evidence that emotionally aroused rats show less alternation than relaxed animals. To get at that issue, Douglas moved his rats vigorously in an up-and-down direction, assuming that he would thereby equally upset them as when the motion occurs in the horizontal plane, but not disorient them spatially. The SAB rate came in at 79.2%. If emotionality were the proper explanation for the results of the manipulations intended to disrupt orientation, surely these up-and-down rats would be more disturbed than those receiving one 360° rotation between trials. In fact, the latter procedure reduced SAB, whereas the former did not. To cap this argument, Douglas conducted yet another experiment in which rats were given eight rapid turns in the horizontal plane between $t_1$ and $t_2$ but tested in the same maze in the same room and with the same paper flooring in place. Given access to the odor-trail cue, these animals alternated at a rate of 66.7%, just about the rate typically found for odor trail alone. So, turning the rats in the horizonal plane seemed to have the effect of wiping out the direction-of-movement cue, while leaving odor-trail operative. If that manipulation had simply frightened the animals, their odor-trail alternation might reasonably be expected to be as much affected as their direction-of-movement alternation.

Based on this impressive array of data, Douglas concluded that for rats in T mazes, there are two and only two cues for SAB. One, an intra-maze cue, is odor trail; the other, an extra-maze, extra-room cue seems to be identified with the direction in which they move as rats traverse the goal arms on $t_1$ and $t_2$. The directionality cue is the more potent, but is also vulnerable to disruption if the animal is disoriented between trials. The two cues add probabilistically, fully accounting for the rate of SAB (approximately 80–85%) when all potential cues are available.

## *Subsequent Studies on Direction of Movement*

Douglas followed up his dissertation with additional investigations, showing, for example, that rats with middle ear disease, which affects the vestibular system, have a reduced rate of SAB to the direction-of-movement cue, though not to odor trail (Douglas, 1966b). Similar results were obtained in mice with congenital

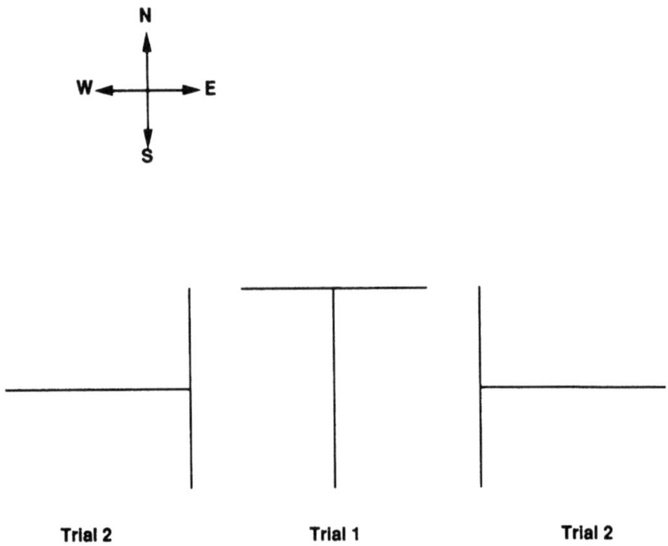


FIGURE 2.6. Maze configurations in the Dember, Sherrick, & Harris (1966) experiment, which revealed rats' tendency to alternate direction of movement from trial 1 starting stem to trial 2 goal arm in the absence of other available cues for SAB.

deficits in otolith development and hence in vestibular function (Douglas, Clark, Erway, Hubbard, & Wright, 1979).

The potency and generality of the direction-of-movement cue were tested in a series of studies in my laboratory, all predicated on the notion that rats leave the maze with a preference for moving in a direction opposite that in which they had most recently been moving. For example, if a rat had been traveling east as it was removed from the maze, it would have for some time afterward a stronger impulse to go west on the next opportunity than to move in any other spatial direction.

Assuming, then, that what matters is the direction in which the animal moves, and not anything about mazes per se, we conducted one experiment (Sherrick & Dember, 1966b) in which $t_1$ was given in a straight alley; $t_2$ occurred in a T maze, with one goal arm of the T pointing in the same direction as the straight alley and the other arm pointing in the opposite direction. On $t_2$ the rats preferred the arm that permitted movement in the direction opposite that enforced by the $t_1$ straight alley. Similar results emerged in a post-hoc analysis of data from a study (Dember, Sherrick, & Harris, 1966) that was designed to investigate odor-trail alternation. That is, two T mazes were set orthogonal to one another; $t_1$ was given in one maze, $t_2$ in the other, thereby neutralizing the direction-of-movement cue (see Figure 2.6). Odor trail was allowed to operate, but the data failed to reveal evidence that odor-trail alternation had actually occurred. What did emerge, however, was evidence that the rats chose the $t_2$ goal arm that pointed in a direction opposite to the $t_1$ starting stem. That unanticipated result may, of course, have been a fluke, but it is intriguing, suggesting that when animals cannot base their effort to alternate directionality as imparted by the $t_1$ goal arms, they may fall back on the direction of the $t_1$ starting stem.

Perhaps most striking were the results of another experiment in this series (Sherrick & Dember, 1966a). Here running speed, not choice, was used as the indicator of the strength of the alternation impulse. Two T mazes were used; the behavior of interest was running speed in the starting stem on $t_2$. The starting stems of the two mazes were either parallel or orthogonal to one another. In the parallel condition, the animals ran either in the same direction on $t_2$ as on $t_1$ (say, both from south to north) or in opposite directions (say, from south to north on $t_1$ and from north to south on $t_2$). With the mazes at right angles to one another, the $t_2$ direction in the starting stem was orthogonal to the $t_1$ direction (for example, south to north on $t_1$ and then east to west on $t_2$). Running speed was fastest when the $t_2$ direction was opposite that on $t_1$ and slowest when the $t_1$ and $t_2$ directions were the same, with the orthogonal conditions intermediate.

Finally, in another series of studies, Sherrick, Brunner, Roth, and Dember (1979) further pursued the logic of this research. In one experiment, rats were allowed to choose one arm of a T maze; they were then run in a straight alley either in the same or in the opposite direction as their T maze goal arm movement. Following this interpolated activity, they were given a standard $t_2$ in the T maze. The interpolated movement either enhanced or reduced $t_2$ SAB, depending on whether it was in the same or opposite direction at the $t_2$ choice. The second experiment in the series showed increasing effectiveness of the interpolated activity as the number of straight alley runs increased. A third experiment involved one run in a straight alley followed by a choice trial in a radial maze. Somewhat surprisingly, only choice of the radial maze arm that enforced movement in the same direction as that in the straight alley was affected: animals avoided that arm, but showed no other preference. That is, they preferred arms that pointed in a *different* direction from that of the straight alley, but showed no special preference for moving in the *opposite* direction. This latter finding is not consistent with the running speed data of Sherrick and Dember (1966a). Clearly, these experiments need to be followed up to determine with greater precision how the direction-of-movement cue operates.

## Subsequent Studies on Odor Trail

Based on a post-hoc analysis of his data, Douglas (1964) had concluded that rats do not alternate with respect to the odor trail of other rats. For his doctoral dissertation, Max Brill (1967) tested that assertion in a more systematic manner. Further, since odor-trail creation and sensitivity to odor trail were likely to interact with gender, Brill used both male and female rats, and the latter both in and out of estrus. Finally, since in the standard SAB paradigm the critical trial ($t_2$) is always the second of a pair of trials, the animals in Brill's experiments were given two trials in the maze, though only their behavior on $t_2$ was of interest. In essence, on $t_2$ rats were offered a choice between two goal arms, one of which contained a paper floor with the odor trail of either another rat or of the target rat. The direction-of-movement cue was neutralized by setting the $t_2$ maze orthogonal to the $t_1$ maze. Several experiments were conducted, yielding results that were quite clear on some questions, but equivocal on others. The following is a fair representation of what Brill found:

1. Male rats alternated to their own odor trail.
2. Anestrus female rats tended to alternate to their own odor trail, though the data are not entirely consistent.
3. Estrus females probably alternated to their own odor trail.
4. Male rats alternated to the odor trail of other male rats, contrary to Douglas's conclusion.
5. Male rats showed a chance level of alternation to the odor trail of female rats, whether the latter were estrus or anestrus.
6. Anestrus female rats failed to alternate to the odor trail of other rats, whether male or female (estrus or anestrus).
7. Estrus females alternated to the odor trail of other estrus females, but not to that of either anestrus females or male rats.

These data have not been replicated, as far as I know, and do not lend themselves to any simple conclusion, except that SAB to odor trail is affected by subject sex and sex of the animal that left the trail. It may be that male rats are attracted to the goal arm where a female had been, if they can make that discrimination, but otherwise avoid the odor trail of another rat, sex unknown, perhaps treating it as their own. The behavior of the estrus females suggests that they treat the odor trail of other estrus females as though it was their own personal trail, while being indifferent or insensitive to the odor trail of anestrus females and males. The general pattern of results suggests that when rats do alternate to odor trail it is because that odor trail is not discriminable from their own, alternating because, they might tell themselves, "I have been here before" not because "another rat has been here before, or is here now."

This last issue arose again, serendipitously, in a study on SAB in gerbils (*Meriones unguiculatus*), which were tested for SAB in the T maze under five conditions, as described by Dember & Kleinman:

"(1) All cues present, (2) only the S's odor trail present, (3) only spatial orientation cues present, (4) neither the S's odor trail nor spatial orientation cues present (to determine if some unknown set of cues might be contributing to the alternation behavior expected to occur under the other conditions) . . . (5) only another S's odor trail present." (Dember & Kleinman, 1973, p. 287)

Condition 5 was included as a follow up to Brill's test of Douglas's assertion about rats' not alternating to the odor trail of other rats. Would male gerbils, as Brill had found for rats, alternate to another subject's odor trail? For reasons now obscure, the 20 male gerbils that served as subjects were housed two to a cage, a detail that proved pertinent to the interpretation of the results.

TABLE 2.1. T maze alternation behavior by gerbils under various cue conditions.

| Condition | Alternation proportion | Number of test trials | $\chi^2$ |
|---|---|---|---|
| 1. All Cues | 0.725 | 40 | 8.10† |
| 2. S's Odor Trail | 0.675 | 40 | 4.90* |
| 3. Spatial Orientation | 0.550 | 40 | 0.40 |
| 4. No Cues | 0.500 | 20 | 0.00 |
| 5. Other S's Odor Trail | 0.525 | 40 | 0.10 |

*$p = .05$
†$p = .01$
Reprinted from Dember and Kleinman (1973).

The results revealed that gerbils did alternate to their own odors trails, but not to direction of movement (see Table 2.1). It also appeared that a chance level of SAB occurred in condition 5; that is, alternation was not evident when another animal's odor trail was present in one of the goal arms. However, when that odor trail was attributable to a noncagemate, SAB rate was 0.70 ($0.10 > p > 0.05$); when the trail-marker was a cagemate, SAB rate was 0.35 ($p > 0.10$). Neither value is significantly different from chance, by conventional standards, nor is the difference between them ($0.10 > p > 0.05$). However, the trend, entirely unexpected, is intriguing and deserves comment. Assuming the results to be reliable, they suggest that gerbils treat the odor trail of noncagemates as they do their own, whereas their familiarity with cagemates enables them to discriminate between self-marked odor trails and other-marked odor trails. If gerbils categorize an odor trail as self-marked, they will alternate to it, even if in fact that odor trail belonged to another (unfamiliar) animal. Should this interpretation prove correct, it implies that gerbils (and perhaps rats) alternate with regard to odor trail not in order to avoid entering an arm where another animal had been, or might now be, but rather to avoid returning to a recently visited arm—that is, alternation occurs in the service of exploration, not out of fear or respect for territorial boundaries. And that brings us to the issue of motivation.

## *What are the Motives?*

### *Cues and Motives*

While the question about the nature of the cues for alternation is separate from that about the motivation for alternation, the two issues are nevertheless intertwined. Thus, for example, the reactive inhibition account of SAB not only identifies the animals' turning response at the choice point as the cue for alternation, it also endows that cue with motivational potency. That is, a temporary aversive state is established whenever a response is made, reducing the probability of the recurrence of that response. The same pertains to Glanzer's stimulus satiation concept: Exposure to a stimulus temporarily reduces the animal's tendency to approach that stimulus again (so, some other stimulus is approached by default).

The close tie between the cues and motives issues is also manifested in the way Douglas and others, myself included, reacted to the evidence that visual cues played no (or at best a minor) role in the regulation of SAB. Such data seemed to call into question the response-to-change explanation. That is, if SAB simply reflects a more general tendency to respond to or for change in stimulation, then why do rats fail to alternate when only visual cues, to which they are apparently sensitive, are present to guide their behavior? Douglas (1964) concluded that SAB and response to change are different phenomena, linked only historically.

Finally, note that in identifying direction of movement and odor trail as the dominant cues for SAB, Douglas was silent on why animals bother to use those cues. So, even if the cues issue were entirely resolved (which it probably is not), there still remains the question of why animals alternate when the critical cues are available. Douglas, in fact, looked toward the brain for an account of SAB; others turned specifically to neurochemical factors implicated in excitatory and inhibitory brain processes. These approaches to the why question are fully covered in Chapter 5 and 6, and need not be explicated here. This chapter will close with a brief look at a few more nearly psychological accounts of why animals alternate.

*Foraging Strategies*

It has been suggested (e.g., Estes & Schoeffler, 1955; Kamil & Sargent, 1981) that SAB represents an efficient foraging strategy. If an animal enters one goal arm of a T maze and finds nothing there to satisfy some biological need, such as food or water, then it would seem adaptive to try the other arm next. Abstractly, this scenario describes a lose-shift strategy.

But recall that SAB occurs in fully sated animals. Moreover, it also occurs in animals that do find some pertinent reinforcement, such as food (Fowler, Blond, & Dember, 1959) in the $t_1$ goal arm or in both goal arms, for that matter. Of course, asymmetrical reinforcement will eventually lead to repetition, as can reinforcement in both goal arms (see Chapter 4 for further details), but in the early trials of most learning experiments, the SAB pattern is quite robust.

Perhaps, then, what SAB reflects is a win-shift strategy. After all, in natural situations, a benevolent experimenter does not conscientiously replenish a cache of food, for example, whenever it is depleted. So the adaptive pattern is to seek elsewhere after reinforcement in one location. But what of those instances when sated animals alternate? And how do we account for those times when a lose-shift strategy seems to apply? Can lose-shift and win-shift strategies both be operative? Perhaps, though one might hope for a more parsimonious conception.

*Walker's Action Decrement*

Mention was made in Chapter 1 of a model proposed by Edward Walker (1956, 1958) based on the notion of a central inhibitory process, the action decrement. That process was meant to supplant the more peripherally focused notions of reactive inhibition and stimulus satiation. What made Walker's model special, however, was the way in which it explicitly tied the action decrement (and hence

SAB) to both reinforcement and the passage of time. Though it has not solved the mystery of motivation, this model did lead to some interesting hypotheses that do not seem to follow from other approaches and to data that are at least partially confirmatory.

In effect, Walker argues that rather than being inimical to learning, the action decrement, in the long run, enhances learning. In the short run, the action decrement inhibits the occurrence of some to-be-learned response (as revealed in SAB); however, the system, which is initially dampened by the action decrement, eventually rebounds, thereby increasing the probability that the behavior in question will be repeated (i.e., "learned"). Working backward from the traditional effect of reinforcement on habit formation, Walker further argues that reinforcement magnitude will be positively related to the extent of the action decrement (and hence to SAB rate when the ITI is relatively short). In sum, any factor that increases the action decrement will also ultimately enhance learning. To test this hypothesis, Walker ran thirsty rats in the standard T-maze alternation situation; half the animals were rewarded with water on $t_1$, regardless of their choice, and half were not so rewarded. All animals were given a second trial in the maze, with ITI varying from about 80 seconds to four hours. In general, the data showed greater SAB for rewarded subjects at the short ITIs and less SAB (greater learning?) when ITI was long. Some additional experiments (see Walker, 1958, for details) also provided encouraging results.

Walker views the action decrement as serving the adaptive function of protecting the consolidation process that underlies habit formation. Looked at in that light, SAB, a behavioral manifestation of the action decrement, is essentially an artifact, without a motive of its own. In that sense, Walker's speculative model has much in common with those approaches that take an even more overtly neurobiological stance on SAB.

## SAB, Response to Change, and Curiosity

In 1957, with Robert Earl, I coauthored an article on curiosity, exploration, and manipulation that sketched a motivational theory built around the notion of complexity. For present purposes the important point was an argument that complexity and novelty are carried by the same psychological processes of confirmation and disconfirmation of expectancies. In that way, stimulus change, as an instance of novelty, became tied to a general theory of curiosity. Response to change, in turn, which heuristically grew out of the SAB paradigm, became a mediating link between SAB and curiosity. And by a bit of verbal maneuvering, the motivation behind SAB was identified as "response for change." That is, the animal alternated in order to optimize the amount of stimulus change (or novelty or complexity) in its current life space. Entering the alternate goal arm was like finishing one novel and then picking up another (rather than rereading the first).

This conception was called into serious question by Douglas's conclusion that visual cues play no role in SAB, even though they obviously do in the response-to-change paradigm. If SAB and response to change are manifestations of a more general curiosity motive, why do they not follow the same rules, vis à vis effective cues?

Two other issues also need to be addressed that, as far as I know,have not previously been raised. (1) How exactly does alternation to direction of movement reflect curiosity? It is easy to see how entering different goal arms, or even going toward different geographical locations, can provide the kind of variation in experience called for by an "optimal stimulation" notion of curiosity. But what does it do for one's curiosity to vary the direction in which one has just moved? Direction of movement per se seems too abstract to serve that function. However, it might be argued that alternating direction of movement, though not stimulation specific, provides a general heuristic device for the animal that assures access to varied stimulation. (2) The same might be said about the tendency to avoid one's odor trail. But here another interesting question comes to mind. Consider the rat in the choice point of a T maze on $t_2$. Direction of movement has been essentially neutralized by setting the $t_2$ T maze orthogonal to the $t_1$ T maze. All that the animal has available to guide its choice is the odor trail if left in one of the goal arms. But that odor trail was not present in that arm on $t_1$. With regard to odor trail, the other arm is as it was on $t_1$. That is, on $t_2$ the animal is faced with a choice between a changed (no trail present to trail present) and nonchanged arm. According to the response-to-change notion it ought to select the changed arm; in so doing, of course, it would reenter the arm it had already visited. Thus, in any condition where odor trail is operative, there may be a conflict between two tendencies, alternation and response to change. That may be why the role of SAB to odor trail alone is so low (0.65 or less). One can, of course, imagine other scenarios. For example, it may be that response to change simply does not pertain to odors or odor trails. After all, its demonstration has been confined thus far to visual stimuli. It may also be that the alternation and response-to-change tendencies are not additive, although data obtained by O'Connell (1964) suggest otherwise, at least when maze-arm brightness is the source of stimulus change.

Finally, one could claim that when rats are sufficiently disoriented through various maze manipulations, whatever visual cues are present are simply not sufficient to guide efficient exploration, especially if those cues are not very salient in the first place (see Richman et al., 1970, who argue just that for visual cues in T-maze goal arms). In short, the animals may not attend to visual cues on $t_1$, then, on $t_2$, when visual cues are all they have to go on, it is too late to use them since the animals have no visual representation of the $t_1$ maze. Absent that representation, there is no basis for systematic choice.

So, curiosity may indeed motivate SAB. How to test that hypothesis empirically is a challenge for future researchers. In the meantime, as will become evident, there is a great deal that has been and can be learned about SAB even in relative ignorance of why animals engage in that behavior.

# 3
# Phylogenetic Comparisons

ROBERT N. HUGHES
UNIVERSITY OF CANTERBURY

In their seminal review of alternation behavior, Dember and Fowler (1958) unashamedly confined their attention to the literature on rats. They correctly reasoned that, at the time, the phenomenon had not been sufficiently investigated in other species to warrant consideration. Thirty years later we must ask whether the situation has improved enough to provide meaningful species comparisons with a view to understanding the evolution of alternation tendencies. Unfortunately, most assessments of species differences still involve comparing data with that already published for the rat or some other animal. There are relatively few studies on record in which two or more species have been compared within a single investigation. Because of insufficient information, the phylogenetic generality of factors known to influence and control rat alternation is difficult to assess. Consequently, a review of "phylogenetic comparisons" is effectively restricted to determinations of the prevalence of alternation in species other than the rat for which, of course, it was first described.

Implicit in many studies of alternation in species other than rats is the assumption that, because of its obvious survival value, the phenomenon must be widespread throughout the animal kingdom. A tendency to sequentially alternate in natural settings should facilitate foraging and exploratory behavior (Richman et al., 1987) and ensure relatively direct paths of movement towards or away from significant stimuli (Hughes, 1967b). These effects would in turn assist dispersal and improve access to food, shelter, mates, and so forth. In this chapter I will attempt to determine how widespread the phenomenon really is. For convenience I will distinguish between more conventional SAB observed from two or more successive trials in a T maze or something similar, and spontaneous alternation of body turns (or turn alternation) involving turning in the opposite direction to a preceding forced turn during the course of a single trial. As described in earlier chapters, SAB in rats mainly involves active choices of external stimuli based on previous experiences and thus familiarity with those stimuli. In contrast, turn alternation, particularly in invertebrates, relies more on feedback from proprioceptive (nonvestibular) cues arising from previous responses (Hughes, 1985; Schäfer, 1983). But in many studies of SAB the extent to which different species can and will alternate either stimuli or responses is not known because of failure to incorporate procedures that enable distinctions to be drawn between stimulus and response factors. One useful procedure is to vary duration of confinement to a goal arm (intra-trail interval): longer time periods should elevate stimulus, but reduce response alternation (Dember, 1961). Whereas alternation of external stimuli seems improbable following a forced turn during a single trial, stimulus or response alternation may be equally likely on successive trials in many species other than the rat for which stimuli in the form of olfactory and spatial cues are most important (Douglas, 1966a). Since studies of rats and to a lesser extent humans, comprise most of the literature on alternation behavior, these two species will not be considered as separate cases in this chapter.

## SAB in Invertebrates

Even though they comprise over 90% of known animal species, there are hardly any studies of SAB in invertebrates on record. In part, this might be due to the practical difficulties of simulating the rat SAB paradigm involving disruptive or even injurious intertrial handling. But techniques exist that would minimize these effects, such as the use of interchangeable goal and start boxes (e.g., Wayner & Zellner, 1958). Alternatively, failures to observe SAB may have gone unreported because there is less interest in the publication of negative findings.

### *Annelids*

The most primitive invertebrates in which SAB has been reported are the segmented oligochaete worms. During consecutive trials in a T maze, Wayner and Zellner (1958) observed significant alternation in a species of earthworm, *Lumbricus terrestris*, which decreased to chance levels after removal of the suprapharyngeal ganglia. Surprisingly, they found no difference in SAB between intertrial intervals of 5 and 10 minutes. Nevertheless, the authors interpreted their results as favoring reliance of the phenomenon on reactive inhibition-based responses rather than on exploration-related stimuli. Although Iwahara and Fujita (1965) later failed to observe SAB in another species of earthworm, *Pheretima communissima*, removal of the suprapharyngeal ganglia produced significant repetition. The conflicting results of the two studies could have arisen from several procedural differences between them or from the different species studied. For example, Fraser (cited in Jacobson, 1963) found that while one species, *Allolobophora terrestris*, alternated, another, *Lumbricus rubellus*, significantly repeated. It would therefore appear that earthworms may either alternate or repeat for reasons not yet understood. The relative involvement of response and stimulus cues in either tendency is not known.

### *Crustaceans and Myriapods*

In a study of left-right preferences, MacKay (1945) noted that during 50 consecutive trials in a Y maze, a species of hermit crab, *Calcinus herbsteii*, consistently turned in the same direction as its large left claw. Another hermit crab,

*Clibnarius zebra*, with claws of equal size, demonstrated the same left-turning bias (MacKay, 1947). Rather than alternation, response repetition (possibly arising from the right spiral structure of the shells they inhabit) prevailed for these species, although no information is available about any role of maze stimuli in their choices.

The only crustaceans in which SAB rather than turning preferences has been investigated are terrestrial isopods. Iwahara (1963) failed to observe T-maze SAB with either 2 or 11 consecutive trials and intertrial intervals from 4 seconds to 12 hours in the pillbug, *Armadillidium vulgare*. I recently supported this finding in an unpublished experiment with a related isopod, *Porcellio scaber* (the wood louse or sowbug). To maximize the likelihood of SAB's occurring, 72 subjects were forced left or right in a Perspex T maze with a minimal intertrial interval but intra-trial intervals of 0, 120, and 600 seconds. On a second free trial the numbers out of 24 that alternated in the three conditions were 13, 10, and 10, respectively. None of these was above chance, nor were differences between them significant. Studies of crabs and isopods have so far failed to yield evidence of SAB in crustaceans.

There appears to be only one investigation of SAB in myriapods (centipedes and millipedes) on record. In this, *Lithobius forficatus* alternated in a T maze provided that opportunities for body turns in the interval between trials 1 and 2 were prevented (Schäfer, 1983). Since variations in visual and tactile characteristics of the goal arms did not affect the tendency, the author concluded that the behavior was response- rather than stimulus-guided.

### *Insects*

In view of the enormous number and variety of species that exist, studies of SAB in insects are disappointingly limited. This seems particularly unfortunate considering their behavioral diversity and the highly developed sensory, motor, and integrative capacities of many species.

In an unnamed species of cockroach, repetition rather than alternation occurred in a Y maze, a tendency that was unaffected by intertrial intervals of 20 and 120 seconds (Iwahara & Soeda, 1957). Repetitive behavior increased when electric shock preceded test trials (Iwahara & Soeda, 1957). By far the most elegant demonstration of invertebrate SAB on record was provided by Wilson and Fowler (1976), who gave cockroaches (*Blatta orientalis*) forced trials followed by free choices in a T maze. They observed SAB with long intra-trial (5 and 10 minutes) and "short" intertrial intervals (0, 15, and 30 minutes), visually discriminable goal arms (e.g., light gray versus black), and following one or more forced trials. This study is notable for establishing reliable SAB in cockroaches and for showing that when the conditions are right, cockroaches are responsive to a number of influences known to affect stimulus alternation in rats (Richman et al., 1987). The apparent exploration-based nature of their SAB is in line with evidence of their exploratory tendencies in other situations (Darchen, 1952, 1955).

Apparently the only other insect for which behavior of some relevance to SAB has been investigated is the parasitic beetle *Aleochara bilineata*. During 20 consecutive trials, members of this species repeated their entries of Y-maze arms, with different individuals showing different arm preferences (Putnam, 1962). It was suggested that their recapture in a vial after each trial may have provided the beetles with a rewarding exploratory environment, thereby reinforcing subsequent choices of the same arm. But alternatively, if the recapture process were aversive, it could be argued that this encouraged repetition through avoidance of the more novel alternate arm in the manner shown for rats following other aversive experiences (Aitken, 1972; Haywood & Wachs, 1967). A similar interpretation might account for the shock-induced enhancement of repetition in cockroaches described earlier in this section (Iwahara & Soeda, 1957).

### *Discussion*

Apart from examples of presumably response-guided alternation in some species of earthworm, a centipede species and stimulus-guided alternation in a cockroach species, there is as yet little evidence of SAB in the majority of invertebrates. In the absence of sufficient research it is too soon to draw conclusions for most other species even when SAB has been shown not to occur. In this case, failures to alternate may reflect inadequacies of the methodology applied in relation to the capacities and lifestyle of the species concerned. Given the exploration-based nature of stimulus alternation (Richman et al., 1987), with sufficient care it should not be surprising to observe the phenomenon in higher invertebrates known to display exploratory tendencies, e.g., ants (Caldwell & Phaup, 1959), wasps (Beusekom, 1948), and honey-bees (Lindauer, 1952). But unfortunately there is no information available on tendencies to alternate for the vast majority of invertebrate groups. Absence of any records at all for such behaviorally interesting groups as the arachnids and molluscs is regrettable particularly since representatives of the latter phylum have been featured as models in other areas of behavioral research, namely, studies of habituation and sensitization in the marine mollusc, *Aplysia* (Castellucci & Kandel, 1976) and visual discrimination learning in *Octopus vulgaris* (Sutherland, Mackintosh, & Mackintosh, 1965).

## *SAB in Vertebrates*

Although SAB has been investigated in a broader range of vertebrates than invertebrates, there are still surprisingly few studies that do not exclusively use rats, mice, or human beings as subjects. Of these, most deal with mammals. The comparative lack of published studies with nonmammals may be due to reasons proposed for the limited invertebrate literature, namely, practical difficulties in testing and failures to publish negative findings.

### *Fish and Reptiles*

Fish with a bony skeleton (teleosts) appear to be the most primitive vertebrate group in which SAB has been investigated. Neiberg, Dale, and Grainger (1970) noted that, in general, blue gouramis (*Trichogaster trichopterus*) did not alternate in a T maze on their first opportunity. However, those individuals that had previously failed to alternate later selected the same goal arm provided it had changed in illumination between trials. Although this apparent stimulus selection by some subjects also involved response repetition, the authors preferred to interpret their findings as evidence of stimulus alternation. Invariant alternation in goldfish (*Carassius auratus*) has been observed provided that the subjects experienced seven or more forced trials to their preferred goal arm beforehand (Fidura & Leberer, 1974). In a study of the acquisition of win-shift feeding, rainbow trout (*Salmo gairdneri*) spontaneously shifted feeding positions within short time periods as well as varying initial positions each day (Hartman, 1982). The level of this tendency was correlated with ability to learn a win-shift strategy in accord with one of the functions suggested for SAB, namely, facilitation of foraging (Richman et al., 1987). Unfortunately it is not possible to determine the relative involvement of response or stimulus factors in either this example of trout foraging or in the behavior described for blue gouramis and goldfish.

There appears to have been only one published study of SAB in reptiles, namely, an attempt to demonstrate responses to brightness change in the freshwater turtle *Chrysemys sp* (Spigel & Ellis, 1965) using a procedure based on that developed by Kivy et al. (1956). Responses to visual change or novelty are generally accepted as a reason for SAB in rats (Dember & Fowler, 1958). Following exposure to two gray or two black Y-maze goal arms, when later confronted with one gray and one black, the turtles always chose the black arm irrespective of whether or not it was the changed alternative. But this does not mean that, under appropriate conditions, the subjects would not alternate or show responses to other forms of stimulus change.

Evidence to date for SAB in fish and reptiles is not conclusive.

### *Birds*

In spite of including up to 116 forced trials and varied goal arm discriminability (black and white versus two gray arms) in their procedure, Hayes and Warren (1963) observed Y-maze repetition rather than SAB in 2- to 6-day-old domestic chicks (*Gallus domesticus*). Slightly older chicks (10–14 days) also failed to alternate choices of the end boxes of a straight runway entered from a central start box (Dutch, 1969). However, if the visual complexity of the walls of each end box differed, the more complex was consistently chosen, thereby suggesting that complexity rather than novelty might have determined the chicks' preferences. Brown (1976) concluded that SAB in a V maze (with a start box in the junction of the arms) was decreased in 10-day-old chicks following an intra-peritoneal injection of atropine. This conclusion implied significant SAB for saline-injected control birds of which 55% were said to have alternated. Since this percentage represents only 11 out of 20 individuals, control subjects obviously responded at chance levels. But only 3 (15%) and 4 (20%) atropine-treated birds alternated meaning that the drug produced significant repetition ($p = 0.002$ and $0.012$, respectively, my calculations) rather than lower SAB.

Cogan, Jones, and Irons (1979) showed that 1- to 3-day-old chicks would alternate turning responses provided they were allowed to join another chick in the goal box of the chosen arm. These authors adopted Glanzer's (1953b) procedure for distinguishing between response and stimulus alternation and tested their subjects in a + maze that was convertible into one of two T mazes with common goal arms (one black, one white) but different opposing stems. Trial 1 was initiated in one stem and then the maze was rotated 180° to enable trial 2 to commence in the opposite stem. Socially rewarded chicks turned in the opposite direction to their last response. Since this apparent turn alternation involved revisiting the same arm on consecutive trials, it is possible that the chicks had merely exhibited a preference for returning to the goal arm associated with social reward. This interpretation would be less acceptable if the two arms had not been as visually discriminable as they were. So, contrary to their assertion, Cogan et al. (1979) may have merely confirmed some earlier indications of repeated goal-arm choices in chicks. It is likely that such choices involved visual stimulus-guided preferences although Hayes and Warren (1963) also found repetition when both goal arms were gray. Whatever the case, there is currently no convincing evidence of conventional SAB in chicks. For comparison with other species, more research with adults is required since the lack of SAB in favor of chance responding or repetition could be due to the very young ages of the birds tested. While failures to alternate have been observed in infants of some mammalian species such as rats (Kirkby, 1967), mice (Kokkinidis, Irwin, & Anisman, 1979), rabbits (Baisden, Isaacson, Woodruff, & Van Hartesveldt, 1972), and cats (Frederickson & Frederickson, 1979), repetition may characterize free-choice strategies at early stages of human development (Gerjuoy & Winters, 1968). Interestingly, both chicks and humans rely heavily on visual cues for dealing with their environment. (For a detailed discussion of the ontogenesis of SAB, see Chapter 7.)

Clearly it is not possible to draw any conclusions about the prevalence of SAB among birds in general from studies of domestic chicks.

### *Marsupials*

Out of a total of 15 opportunities (3 per subject), 5 adult Virginian opossums (*Didelphus virginiana*) alternated 10 times in a T maze (Tilley, Doolittle, & Mason, 1966). After allowing for initial goal-arm preferences, the authors concluded that this proportion (67%) represented significant SAB, though that may be questioned. Related to this conclusion was a report of responses to brightness change by adult opossums in a Y maze (Platt & James, 1967) in the manner shown for rats in T mazes by Dember (1956). While this finding demonstrated responsiveness to an established determinant of SAB in rats, namely, visual novelty (Dember & Fowler, 1958), it was not confirmed in a second study with 90-day-old opossums in a T maze (Platt, Yaksh, & James, 1968). These authors accounted for the discrepancy by maturational and arousal differences between subjects, although different types of apparatus were used in the two experiments.

The evidence for SAB in at least Virginia opossums seems promising, particularly since several other Australian marsupials have shown varying degrees of positive responsiveness to novel objects (Kirkby & Preston, 1972; Russell & Pearce, 1971).

*Lagomorphs and Rodents*

It is not until placental mammals are considered that definitive evidence of SAB in species other than rats and humans begins to emerge. This may be partly due to more placentals having been investigated since many of them are better suited to rat-type experimental procedures than other species.

While infant rabbits did not alternate unless treated with the anticholinesterase drug physostigmine (Baisden et al., 1972), adult gray rabbits (*Oryctolagus cuniculus*) exhibited about 88% SAB in a Y maze (Hughes, 1973). This rate was comparable to the estimated population mean of 85% for rats in T mazes (Douglas, Mitchell, & Kentala, 1972). Among rodent species other than rats, there have been reports of SAB in mice, guinea pigs, gerbils, and hamsters. Mice (*Mus musculus*) of Quackenbush and New Zealand White strains alternated in a T maze about 70–75% of the time when intertrial intervals were 30 seconds or less (Blampied & Wilby, 1975; Petchkovsky & Kirkby, 1970), with females alternating slightly less often than males (Blampied & Harvey, 1979). This frequency was marginally higher than the 60–70% free-running alternation rate in a Y maze reported for A/J, DBA/2J, and C57BL/6J inbred strains (Anisman & Kokkinidis, 1975) and Swiss Webster mice (Kokkinidis & Anisman, 1976a, 1976b). However, Syme and Syme (1977) observed about 65% conventional T-maze SAB in Simpson inbred mice, a rate that decreased with intra-trial intervals of 30 minutes or longer. In general, SAB rates in mice were lower than for rats in T mazes, although Syme and Syme (1977) observed 88% in a BALB/C strain. As is the case for rats, extended intra-trial intervals (Glanzer, 1953b) and gender (Hughes & Daley, 1977) can influence SAB in mice, for which genetic strain is also important (Henderson, 1970; Syme & Syme, 1977).

Guinea pigs (*Cavia sp.*) older than 10 days have demonstrated about 90% SAB in a T maze (Douglas, Peterson, & Douglas, 1973) as well as freely entering a novel alley on about 80% of their opportunities (Johns, Louis, Becker, & Means, 1982). When tested in a T maze for response versus stimulus alternation with Glanzer's (1953b) procedure, Mongolian gerbils (*Meriones unguiculates*) were said to have exhibited stimulus-guided SAB (Schnurr, 1971). But in a later study, Greenberg (1973) failed to observe SAB in gerbils tested in a T, U, or parallel-arm maze. This finding was challenged by Dember and Kleinman (1973), for whom gerbils alternated in a T maze provided that objects on which to leave odor-trail cues were glued to the maze floor. Greenberg (1978) responded with a series of five experiments in which the subjects were able to leave odor cues on objects. Again there was no evidence of alternation in this species. Although the case for SAB in gerbils is not proven, it seems unlikely that they would not alternate in view of their high levels of exploratory behavior compared with rats (Thompson & Lippman, 1972).

Sinclair and Bender (1978) concluded that hamsters (*Mesocricetus auratus*) do not alternate even though Kirkby and Lackey (1968) had earlier reported about 93% T-maze SAB in this species. Recently, I observed 80–85% stimulus-guided SAB in hamsters on their first opportunity in a T maze provided that intra-trial intervals of 30 seconds were experienced (Hughes, 1988). This rate dropped to 61% when SAB over eight successive trials was considered. Longer intra-trial intervals increased alternation for males only, whereas varied goal-arm discriminability (black and white versus unpainted arms) had no effect on either sex, thereby suggesting that nonvisual (possibly olfactory) cues were involved. Depending on experimental conditions, it is likely that hamsters alternate at a rate similar to or even higher than that of rats, which they resemble both with regard to the effects of intra-trial interval (Glanzer, 1953b) and goal-arm discriminability (Richman et al., 1987).

*Carnivores*

While being unresponsive to increased goal-arm discriminability, domestic cats (*Felis domesticus*), 5 weeks of age and older, showed SAB in a T maze, with adults alternating invariantly (Frederickson & Frederickson, 1979). When given a single opportunity in a Y maze, ferrets (*Mustela putorius*) displayed 95% SAB (Hughes, 1965). This rate dropped to about 66% over several trials but increased again to 79% with a 2-minute intra-trial interval (Hughes, 1967a) and an intraperitoneal injection of saline solution (Hughes & Greig, 1975). When comparing different species, these results suggest a need to distinguish between SAB rates derived from observations of a single pair of trials and those involving several trials. It therefore seems likely that, contrary to an earlier assertion (Kirkby & Lackey, 1968), ferrets may alternate less than rats except perhaps on their first opportunity. Both carnivores appear affected by goal-arm discriminability (in the case of cats) and intra-trial interval (in the case of ferrets) in a similar way to rats, with ferrets also showing rat-like responsiveness to brightness change and complexity in SAB settings (Eastment & Hughes, 1968; Hughes, 1964, 1965).

*Discussion*

Although encouraging, the evidence for SAB in vertebrates apart from rats and human beings is not overwhelming. This situation reflects the lack of sufficient studies of other species rather than failures to observe the phenomenon. As with invertebrates, there are significant vertebrate groups that have not yet been studied. We know nothing of SAB in the vast majority of fish, amphibians, reptiles, and birds. Among mammals, there are no studies of several important groups such as insectivores, cetaceans, ungulates, and ruminants. We do not even know if SAB occurs in many of our more familiar domestic animals such as pigs, goats, sheep, cows, horses, and dogs, not to mention more exotic mammals encountered in the wild or in zoos. Apart from humans (Schultz, 1964) and a demonstration of repetition rather than alternation in rhesus monkeys (Hosoda, 1963), little is known about SAB in primates even though monkeys featured extensively in some of the earlier studies of exploratory behavior (e.g., Butler, 1953; Harlow, 1950).

From the limited evidence, there seems little doubt that SAB typifies some fish, Virginia opossums, rabbits, mice, guinea pigs, hamsters, cats, ferrets, and of course rats and humans. Whereas domestic chicks appear to lack the tendency, gerbils probably can alternate even though evidence for and against is still evenly balanced. Of the species shown to alternate, goldfish, guinea pigs, and cats appear to do so at higher rates than rats. Opossums and most mouse strains alternate less than rats. While rabbits appear similar to rats in SAB rate, hamsters and ferrets may alternate more or less depending upon the procedure adopted. Since there does not seem to be any obvious phylogenetic trend to account for these species differences in SAB rate, it may be more fruitful to consider the phenomenon's evolution in terms of the concept of anagenesis, i.e., the ranking of behavioral capacities rather than different kinds of animal (Gottlieb, 1984). Such a ranking, while often at variance with a phylogenetic ordering, could be a more sensible way of viewing behavioral evolution and accounting for apparent similarities between widely separated species. However, rather than anagenetic trends, it is more likely that failures to alternate and species differences in SAB rates reported so far are due to inadequately standardized procedures interacting with poorly controlled subject variables.

Both procedural and subject details are already known to influence SAB in rats (Dember & Fowler, 1958; Richman et al., 1987). Some of the more important details include the subjects' developmental status and emotionality, their dominant sense, and their sensory and memorial capacities plus their ability to process and habituate to stimuli encountered on successive SAB opportunities. Whether a T, Y, or some other type of maze is used might be important, as could goal-arm dimensions in relation to the size, locomotor style, and stimulus-processing speed of the animal. Related to this is the subject's life-style—aquatic, ground-living, arboreal, and flying species might respond differently to the demands of conventional SAB settings. With such settings, initial goal-arm preferences should be taken into account when estimating whether or not an observed SAB level differs from that expected by chance (Dember & Fowler, 1958; Douglas, 1966a, Manning, 1973a).

Perhaps the habit of defining a choice as a complete entry of a goal arm might be too rigorous for some species. I have found that, although arm entries by rats may suggest the ineffectiveness of an independent variable, different conclusions arise if the arm first approached but not necessarily entered is used as an index of SAB (Hughes, 1966b). In other words it might be useful to distinguish between noticing the alternate arm on trial 2 (attention) and choosing to enter it (preference). In arriving at an estimate of SAB rate for a species, distinctions should also be made between the provision of single and multiple alternation opportunities for individuals, particularly as the tendency is known to decline in rats with greater numbers of successive trials (Dember & Fowler, 1958). Of the studies I have discussed, very few have recognized the need to consider these procedural and subject variables. Nor have there been many serious attempts to investigate either within or between species those influences known to modify SAB in rats, outlined by Dember and Fowler (1958) and Richman et al. (1987). A number of these might be useful in determining the contribution to SAB (or its lack) of some of the subject variables mentioned earlier. For example, varying intertrial intervals could be useful in assessing memorial capacities, whereas stimulus-processing ability might be reflected in the effects of different intra-trial intervals. As suggested for invertebrates, this latter manipulation could also help determine if SAB were mainly response- or stimulus-guided, a question rarely addressed in species other than the rat.

Even when SAB is assumed to be stimulus-guided, there have been few attempts to determine in different species the relative involvement of visual, olfactory, or some other form of stimulation. Rather, the emphasis has remained on establishing whether or not the phenomenon exists. Consequently, it is not possible to decide if the mechanisms controlling rat SAB also apply to other species. Although some animals can react to visual cues known to affect rat SAB, this does not mean that such cues are normally crucial for the response. They are certainly less important for rats than was once believed (Douglas, 1966a).

Clearly there is now more evidence of SAB in species other than rats and humans compared with 30 years ago. But it is disappointing that the evidence is not stronger. Nevertheless, since alternation is similarly affected by several variables in rats and such widely separated species as cockroaches (Wilson & Fowler, 1976) and ferrets (Hughes, 1967a), it seems likely that further careful experimentation with many more species should consolidate the phylogenetic generality of stimulus-guided SAB.

## *Turn Alternation in Invertebrates*

Although interest in alternation of body turns during single trials originated in studies with rats (Dashiell & Bayroff, 1931), during the last 30 years the phenomenon has become more closely identified with invertebrates. At one time turn alternation was regarded as synonymous with SAB in rats. But it is now clear that turn alternation involves mainly response-generated cues rather than the active choices between currently and previously experienced external stimuli that characterize most examples of SAB (Hughes, 1985; Richman et al., 1987). In most demonstrations of turn alternation, a forced turn is followed some distance later by either a T junction (where only a 90° left or right turn is possible) or an opening allowing a free turn of any size in either direction. The advantage of the latter procedure is that, in addition to indicating whether or not alternation has occurred, the magnitude of a turn can also be measured.

Twenty years ago, Lester (1968) reviewed the evidence for body-turn or response alternation in different animal species and concluded that the phenomenon had consistently occurred in only two species, namely, bugs and human beings. It is unclear if the term "bug" was meant to include crustacean sowbugs along with several insects that were described. If not, then attention should be drawn to other pre-1968 examples of turn alternation in isopod crustaceans not mentioned by Lester (1968), namely, those documented by Hughes (1966a, 1967b), Iwata and Watanabe (1957a, 1957b), Kupfermann (1966), and Watanabe and Iwata (1956). Lester also overlooked a very significant series of experiments dealing with the tropical millipede, *Trigoniulus lumbricinus* (Barnwell, 1965). Several more recent studies have confirmed the existence of turn alternation in millipedes and other related myriapods (e.g., Burger, 1971; Mittelstaedt, Mittelstaedt, & Mohren, 1979; Schäfer, 1972, 1982b).

Lester (1968) justifiably concluded that, at that time, evidence favoring turn alternation in protozoan paramecia, turbellarian flatworms, and annelid earthworms was conflicting. Today the situation has changed a little, since further information has appeared for the latter two groups. In the case of turbellarians, it is noteworthy that Shinkman and Hertzler (1964) had reported significant alternation in *Dugesia tigrina* following three or more consecutive 90° forced turns. However, Rice and Lawless (1957) who had earlier failed to observe alternation in *Planaria dorotocephala* applied no more than two forced turns. Perhaps more forcing is required to demonstrate the phenomenon in flatworms (and possibly other species). More recently Aderman and Dawson (1970) reported that two turbellarian species (*Dugesia dorotocephala* and *D. tigrina*) alternated following five forced turns. I also recently demonstrated turn alternation in the earthworm *Lumbricus rubellus*, which was unrelated to the size of a preceding forced turn (Hughes, 1987).

Because fruit flies (*Drosophila melanogaster*) showed sequential repetition when confronted with a series of T choice points (Murphey, 1965), Lester (1968) concluded that this species does not alternate. However, using a more conventional procedure, May and Wellman (1968) later observed alternation at a single choice point in fruit flies following a forced turn. More recently Bicker and Spatz (1976) saw them sequentially alternate in a multiple-T maze when forced turns were interposed between a first and second choice point. From these results fruit flies appear to behave in a similar fashion to other insects when forced turns are experienced prior to a free choice. A notable exception to this rule for insects might be the ant species *Formica subintegra* and *Camponotus novaboracensis*, which did not alternate (Dingle, 1962) in spite of Schneirla's (1929) earlier positive finding.

In addition to the preceding examples, turn alternation has been observed in several other invertebrate species since Lester's (1968) review: the isopod *Oniscus asellus* (Schäfer, 1982a, 1982b); the millipedes *Schizophyllum sabulosum* (Burger, 1971), *Spirostreptus sp.* (Mittelstaedt, Mittelstaedt, & Mohren, 1979), *Ommatoiulus varius* (Schäfer, 1982a) and *Ophiulus pilosus* (Hughes, 1987); the centipedes *Lithobius forficatus* (Schäfer, 1972) and *Scolopendra cingulata* (Schäfer, 1982a); the cockroach *Blatella germanica* and stick insect *Carausius morosus* (Mittelstaedt-Burger, 1972); the larval beetle *Zophobas rugipes* (Wilson & Hoy, 1968); the milkweed bug *Lygaeus kalmii* (Aller,Hirai, & Caldwell, 1979); the earwig *Forficula auricularia* (Hughes, 1987); and the scorpion *Euscorpius italicus* and spider *Clubiona holosericea* (Schäfer, 1982a). Although impressive, it is surprising that this list is not considerably longer given the enormous number of invertebrates in existence. Perhaps the study of turn alternation may have become less "fashionable," although the work of Schäfer and others in Germany show that it is still alive and well in that part of the world. Another possibility might reflect the view that a single underlying mechanism will account for turn alternation in all invertebrates thereby dispensing with any need to extend the range of species studied. This view is now deemed inappropriate, since it is likely that different mechanisms might characterize different species or even the same species at different times (Hughes, 1987).

Contrary to what has typified SAB, most studies of invertebrate turn alternation have usually included assessments of the effects of variables known to modify the phenomenon, not merely establishing its existence. The most commonly investigated variables are pre- and post-forced-turn distances, numbers of consecutive forced turns in one direction, and the angle of a forced turn. The effects of varying these in different species are summarized in Tables 3.1–3.4. Contributions of the eyes and antennae to turn alternation are also featured in a number of studies.

### *Start to Forced-Turn and Forced-Turn to Choice-Point Distances*

It is evident from Table 3.1 that, whereas increasing the distance between the start and forced turn had no effect on the number or size of alternating turns in several isopod and myriapod species, boxelder (*Leptocoris trivittatus*) and milkweed bugs (*Dysdercus* and *Oncopeltus fasciatus*) and larval (*Tenebrio molitor*) and adult beetles (*Tetraopes tetraophthalmus*) showed consistent increases in the tendency. When the distance was increased between a forced turn and choice point, alternation decreased in all isopod, myriapod, and insect species with the exception of *A. vulgare* (Table 3.2). Although no effect of this manipulation was observed with the flatworm *P. dorotocephala*, Aderman and Dawson (1970) noted that alternation increased in *Dugesia dorotocaphala* and *D. tigrina* when distances between five consecutive forced turns progressively decreased.

### *Number and Angle of Forced Turns*

While two species of *Paramecium* and the turbellarian flatworm *P. dorotocephala* were unaffected by greater numbers of forced turns in one direction, another flatworm, *D. tigrina*, showed increased alternation with three or more turns (see TAble 3.3). A species of earthworm (*Lumbricus rubellus*) and several isopods, myriapods, and insects all showed more alternation with increased numbers of forced turns. Of the 22 studies outlined in Table 3.4 in which the size of a forced-turn angle was varied, increases in alternation accompanied sharper angles in 19 involving isopods, myriapods, insects, and arachnids.

TABLE 3.1. Effects on turn alternation of increased start-to-forced-turn distance.

| Species | Class (subclass or order) | Distances in cm | Effect | Author (date) |
|---|---|---|---|---|
| *Armadillidium vulgare* | Crustacea (isopoda) | 3, 15 | None* | Kupfermann (1966) |
| *Porcellio scaber* | Crustacea (isopoda) | 3–12 | None | Hughes (1966a, 1967b) |
| *Oniscus asellus* | Crustacea (isopoda) | 2–12 | None* | Schäfer (1982a) |
| *Lithobius forficatus* | Myriapoda (diplopoda) | 4–16 | None* | Schäfer (1972, 1975a) |
| *Tetraopes tetraophthalmus* | Insecta (coleoptera) | 5, 10 | Increase | Akre (1964) |
| *Tenebrio molitor* (larva) | Insecta (coleoptera) | 3–12 | Increase | Dingle (1964a) |
| *Leptocoris trivittatus* | Insecta (hemiptera) | 0–20 | Increase | Dingle (1961, 1964b) |
| *Dysdercus fasciatus* | Insecta (heteroptera) | 0–20 | Increase* | Dingle (1965) |
| *Oncopeltus fasciatus* | Insecta (heteroptera) | 0–20 | Increase* | Dingle (1965) |

*Angle turned

TABLE 3.2. Effects on turn alternation of increased forced-turn-to-choice-point distance.

| Species | Class (subclass or order) | Distances in cm | Effect | Author (date) |
|---|---|---|---|---|
| *Planaria dorotocephala* | Turbellaria (tricladida) | 2–18 | None | Rice & Lawless (1957) |
| *Armadillidium vulgare* | Crustacea (isopoda) | 5–20 | None* | Kupfermann (1966) |
| *Porcellio scaber* | Crustacea (isopoda) | 3–13.8 | Decrease | Hughes (1966a, 1967b) |
| *Oniscus asellus* | Crustacea (isopoda) | 2–12 | Decrease* | Schäfer (1982a) |
| *Lithobius forficatus* | Myriapoda (chilopoda) | 4–16 | Decrease* | Schäfer (1972, 1975a) |
| *Tetraopes tetraophthalmus* | Insecta (coleoptera) | 5–9 | Decrease | Akre (1964) |
| *Tenebrio molitor* (larva) | Insecta (coleoptera) | 1.5–12 | Decrease* | Dingle (1964a), Grosslight & Harrison (1961), Grosslight & Ticknor (1953) |
| *Leptocoris trivittatus* | Insecta (hemiptera) | 1.25–12.5 | Decrease | Dingle (1961) |

*Angle turned

TABLE 3.3. Effects on turn alternation of increased numbers of forced turns in one direction

| Species | Class (subclass or order) | Number of turns | Effect | Author (date) |
|---|---|---|---|---|
| *Paramecium micronucleatum* | Ciliophora (ciliata) | 0–2 | None | Lepley & Rice (1952) |
| *P. caudatum* | Ciliophora (ciliata) | 0–9 | None | Lachman & Havlena (1962) |
| *Planaria dorotocephala* | Turbellaria (tricladida) | 0–2 | None | Rice & Lawless (1957) |
| *Dugesia tigrina* | Turbellaria (tricladida) | 0–5 | Increase (with 3 or more) | Shinkman & Hertzler (1964) |
| *Lumbricus rubellus* | Oligochaeta (terricolae) | 1–3 | Increase* | Hughes (1987) |
| *Armadilladium vulgare* | Crustacea (isopoda) | 1, 2 | Increase | Watanabe & Iwata (1956) |
| *Porcellio scaber* | Crustacea (isopoda) | 0–40 | Increase | Hughes (1967b), Sachs, Klopfer, & Morrow (1965) |
| *Porcellio scaber* | Crustacea (isopoda) | 1–3 | Increase* | Hughes (1985, 1987) |
| *Lithobius forficatus* | Myriapoda (chilopoda) | 1–3 | Increase* | Schäfer (1980) |
| *Ophiulus pilosus* | Myriapoda (diplopoda) | 1–3 | Increase* | Hughes (1987) |
| *Forficula auricularia* | Insecta (dermaptera) | 1–3 | Increase* | Hughes (1987) |
| *Tenebrio molitor* (larva) | Insecta (coleoptera) | 0–2 | Increase | Grosslight & Ticknor (1953) |

*Angle turned

### *Contributions of Eyes and Antennae*

Dingle (1964b, 1965) reported that painting over the eyes of the insects *Leptocoris trivittatus* and *Dysdercus fasciatus* reduced the angle of the their alternating turns. This effect was greatest when the lower rather than upper halves of the eyes were painted. Alternation of intact specimens of *D. fasciatus* was also reduced when testing occurred in red light (Dingle, 1965). Blinding and testing in infrared light entirely eliminated alternation in the related insect *Oncopeltus fasciatus*, whereas infrared light had no effect on larvae of the beetle *Zophobas rugipes* (Wilson & Hoy, 1968). Schäfer (1975b) also failed to observe any effect of blinding or red light on the centipede *Lithobius forficatus*. Although blinding had no effect on alternation after a forced turn in pillbugs (Kupfermann, 1966), both uni- and bilateral blinding decreased sequential alternation by the related isopod, *Porcellio scaber*, in a multiple-T maze (Hughes, 1978). I ascribed this latter result to slower running speeds and thus longer periods of elapsed time

TABLE 3.4. Effects on turn alternation of increased forced-turn angle

| Species | Class (subclass or order) | Angles in degrees | Effect | Author (date) |
|---|---|---|---|---|
| *Lumbricus rubellus* | Oligochaeta (terricolae) | 45–135 | None* | Hughes (1987) |
| *Armadillidium vulgare* | Crustacea (isopoda) | 30–110 | Increase* | Kupfermann (1966) |
| *Porcellio scaber* | Crustacea (isopoda) | 0–135 | Increase* | Hughes (1985, 1987), Schäfer (1986) |
| *Oniscus asellus* | Crustacea (isopoda) | 0–105 | Increase* | Schäfer (1982a, 1982b, 1986) |
| *Lithobius forficatus* | Myriapoda (chilopoda) | 0–120 | Increase* | Schäfer (1972, 1975a, 1982a, 1983) |
| *Scolopendra cingulata* | Myriapoda (chilopoda) | 0–105 | Increase* | Schäfer (1982b) |
| *Trigoniulus lumbricinus* | Myriapoda (diplopoda) | 0–120 | Increase* | Barnwell (1965) |
| *Ommatoiulus varius* | Myriapoda (diplopoda) | 0–105 | Increase* | Schäfer (1982b) |
| *Ophiulus pilosus* | Myriapoda (diplopoda) | 45–135 | None* | Hughes (1987) |
| *Forficula auricularia* | Insecta (dermaptera) | 45–135 | None* | Hughes (1987) |
| *Tenebrio molitor* (larva and adult) | Insecta (coleoptera) | 0–120 | Increase* | Schäfer (1982b), Schäfer & Wegner (1974) |
| *Leptocoris trivittatus* | Insecta (hemiptera) | 0–90 | Increase | Dingle (1964b) |
| *Euscorpius italicus* | Arachnida (scorpion-idea) | 0–105 | Increase* | Schäfer (1982b) |
| *Clubiona holosericea* | Arachnida (araneida) | 0–105 | Increase* | Schäfer (1982b) |

*Angle turned

between consecutive turns, which I had earlier found to be responsible for decreases in alternation with greater distances between forced and free turns (Hughes, 1967b). However, this apparent independence of distance and time is not supported by studies with other species (Dingle, 1964a; Schäfer, 1975a, 1982a). It appears that wide species differences characterize the extent to which visual input can modify turn alternation.

Neither uni- nor bilateral removal of antennae in isopods had any effect on sequential alternation in a multiple-T maze (Hughes, 1978) or on the angle of an alternating turn (Hughes, 1985; Kupfermann, 1966; Schäfer, 1986). Schäfer (1975a) also failed to observed any effect of bilateral antennectomy on the angle turned by centipedes (*Lithobius forficatus*), but, following unilateral removal, noted sharper turns in the direction of the remaining antenna. This latter effect also determined the extent to which centipedes turned in the direction of a followed wall after leaving a straight runway (Schäfer, 1972). It therefore seems likely that involvement of the antennae and presumably tactile input in turn alternation varies with different species.

*Functions and Mechanisms*

When allowed successive choices in a multiple-T maze with few or no forced trials, wood lice, centipedes, and fruit flies exhibit patterns of alternating turns (Bicker & Spatz, 1976; Hughes, 1967b, 1978; Schäfer, 1976b). Through providing a correcting influence (Barnwell, 1965; Dingle, 1961, 1964a; Mittelstaedt-Burger, 1972), sequential alternation facilitates efficient movement towards or away from significant environmental stimuli in the absence of directional cues (Hughes, 1967b, 1978). This function for turn alternation is supported by increases in wood lice following exposure to biological hazards such as dessication (Hughes, 1967b) and starvation (Hughes, 1978). Under such circumstances, greater tendencies to alternate would obviously be advantageous.

Suggested mechanisms for invertebrate turn alternation that have received the most attention are: (1) Hull's (1943) concept of reactive inhibition, and (2) centrifugal swing combined with thigmotaxis. Reactive inhibition advocates essentially maintain that, following a forced turn, inhibition for turning in that direction is established so that turning in the opposite direction occurs when the opportunity arises (Lepley & Rice, 1952). This inhibition dissipates with time or distance traveled and is enhanced by greater numbers of forced turns in the same direction (Grosslight & Ticknor, 1953). According to the centrifugal swing explanation, on encountering a forced turn, each subject is carried forward to some extent by its own momentum and is swung out against the far wall of the apparatus (Schneirla, 1929), which it then follows because of a thigmotactic tendency to remain in contact with vertical surfaces (Jensen, 1959). On reaching a choice point, alternation inevitably occurs whên the subject attempts to continue following the now absent wall by turning in its direction.

Increases in turn alternation following greater numbers and sharper angles of forced turns and decreases with longer distances between a forced turn and a choice point outlined in Table 3.2–3.4 are consistent with a reactive-inhibition interpretation. Centrifugal swing plus thigmotaxis seemed a less likely mechanism since various species still alternated when both those influences were minimized through the use of V-shaped runways and floating mazes without walls (Dingle, 1961, 1964a; Grosslight & Harrison, 1961; Kupfermann, 1966; Schäfer, 1975b). Even though tactile cues are not crucial for alternation in many species, their influence should not be underestimated, particularly when considering the direction turned either without any preceding forced turn or when a vertical surface has been closely followed immediately prior to a choice point. For example, both centipedes and wood lice will turn in the direction of a followed wall when emerging from a straight runway (Hughes, 1987; Schäfer, 1972). After having been forced three times in one direction and then once in the opposite, most wood lice will subsequently alternate with respect to the initial turns – the last turn does not appear to cancel out cumulative effects of the preceding three (Hughes, 1985, 1987). However, after the fourth turn, subjects that came into very close contact with the outer wall of the apparatus later turned in the same direction as this wall (Hughes, 1987). This suggested that, for such individuals, a thigmotactic wall-following tendency may have overpowered any influence of the three preceding turns. When applied to some other species, the procedure outlined revealed that earthworms and millipedes may be more affected by tactile cues than either wood lice or earwigs (Hughes, 1987).

Most of the evidence favors some type of response-generated mechanism for turn alternation in invertebrates with possibly tactile and (as suggested earlier) visual cues being involved to varying degrees for different species. It now seems unlikely that such a mechanism will prove to be reactive inhibition of the type suggested earlier (Lepley & Rice, 1952). The apparent independence of distance traveled and elapsed time between consecutive turns is inconsistent with a reactive inhibition interpretation (Dingle, 1961, 1964a). Likewise, increased alternation in some insects following greater distances between the start and a forced turn outlined in Table 3.1 can not be accounted for by reactive inhibition. In such cases it is possible that longer start alleys may have provided more time for tactile or visual stimuli to exert an additional effect. In an unpublished study, I noted that, as they approached the end of a straight runway, pillbugs moved closer to the followed wall. On emerging they also turned at sharper angles than wood lice, which had not shown this change in distance from a followed wall. Since the concept of reactive inhibition as applied to turn alternation stresses nonrepetition of forced turns, invertebrates should prefer to walk straight ahead rather than alternate when given the opportunity to do either. Quite clearly, alternation is what happens.

More recent research has focused on proprioceptive feedback arising from the negotiation of a turn rather than the higher order inhibitory or memorial processes suggested by some authors (Kupfermann, 1966; Lepley & Rice, 1952). I suggested for wood lice that differences in activity of legs on the two sides of the body might be responsible (Hughes, 1967b, 1978). While turning, legs on the opposite side to the turn would work harder and travel further than those on the same or inner side. Consequently the relatively "rested" inner legs might be likely to have more influence on later turning behavior, thereby encouraging a turn in the opposite direction to a forced turn. Evidence for the effects of such bilaterally asymmetrical leg movements (BALM) was later found in a positive relationship between closeness to the outer wall of a post-turn runway and the size of subsequent alternating turns (Hughes, 1985). It was reasoned that, after a forced turn, the inner legs had driven each wood louse closer to the outer wall, thereby enabling BALM effects to be observed even before a free turn had occurred. My current research involving leg amputations is indicating that such effects may depend on the activity of specific legs particularly near the posterior end. While I earlier suggested that BALM effects could have either a neural or fatigue basis (Hughes, 1985), on reflection the latter possibility seems unlikely. It would be maladaptive for the large bilateral differences in fatigue required for alternation to be generated by a single turn, as well as improbable that such differences could so rapidly dissipate with continued effort, as suggested by the relationship between alternation and forced turn to choice-point distance outlined in Table 3.2.

Instead, it is more likely that isopod alternation arises from proprioceptive feedback from particular legs to sites in the CNS responsible for the integration of inhibitory and excitatory influences on turning behavior.

An alternative explanation for invertebrate alternation involves feedback from the distortion of body segments experienced while turning (Burger, 1971; Schäfer, 1975b, 1982a). However, in a recent unpublished study, I showed that requiring wood lice to negotiate a forced turn that imposed varying demands on the legs but kept body distortions constant, produced a level of alternation consistent with a BALM interpretation. Beale and Wester (1971) had also reported that increasing the activity of legs on one side only, caused wood lice subsequently to turn in that direction. Although Schäfer (1975b) found that amputating different pairs of legs in centipedes had negligible effects on their turn alternation, in terms of a BALM interpretation and my current research, asymmetrical amputations would be required for definitive conclusions.

I am reasonably convinced that, for isopods at least, turn alternation arises primarily from BALM effects. This of course does not preclude the operation of other influences such as body distortions or tactile cues either for isopods or other species under different conditions. As suggested by Dingle (1965), turn alternation in invertebrates may best be regarded as behavior guided by both proprioceptive and external cues although I would weight the former much more heavily than the latter. If BALM influences are important for animals other than isopods, necessary conditions should be a bilaterally symmetrical body and the presence of legs or their equivalent.

## *Turn Alternation in Vertebrates*

According to Lester (1968), at that time, the only vertebrate that had consistently shown response if not body-turn alternation was the human being. He concluded that domestic chicks did not alternate and that the evidence for rats was conflicting. In reaching this latter conclusion, Lester overlooked several earlier studies with rats that demonstrated both sequential alternation at successive choice points and alternation following a forced turn (Ballachey & Buel, 1934; Dashiell & Bayroff, 1931; Witkin & Schneirla, 1937; Zeaman & Angell, 1953). He also failed to mention a report of turn alternation in Japanese quails (*Cotornix cotornix*) after their first two trials (Thompson, Estell, & Loomis, 1965). The case for turn alternation in rats and birds was much stronger in 1968 than suggested by Lester. Since then, alternation following a forced turn has been observed in four species of coral fish (*Thalassoma lutescens*, *Pomacentrus sufflavus*, *Abudelduf uniocellatus*, *Banded dascyllus*; Bate & Kirkby, 1977) and goldfish (*Carassius auratus*; Johnson, 1980) but not in the movement patterns of free-swimming goldfish (Kleerekoper et al., 1970; Westlake & Kleerekoper, 1970). Positive relationships between the magnitude of forced- and free-turn angles have also been reported for two avian species, namely, the Chinese painted quail *Excalfactoria chinensis* (Merkel, 1978; Merkel & Fischer-Klein, 1973) and the savannah sparrow *Passeculus sandwichensis* (Moore, 1982). However, Schäfer (1976a) failed to observed any evidence of turn alternation in mice learning to escape from a T maze.

Although less attention has been devoted to mechanisms for turn alternation in vertebrates than in invertebrates, it has generally been assumed that similar processes are involved for both groups. But recent work on leg movements and body distortions suggest that invertebrate alternation is more appropriately associated with the mechanics of turning behavior. This is difficult to reconcile with alternation in vertebrates arising from higher-order response strategies that might determine such complex activities as navigation in migratory birds. Information about previous turns is obviously stored and then later used to determine the direction and size of subsequent turns. How and at what level this information is stored by vertebrates remains to be determined. Irrespective of the mechanism, turn alternation ensures similar outcomes for both invertebrates and vertebrates, namely, compensation for deviations from a particular path of movement.

## *General Conclusion*

Even though more species are now known to exhibit SAB and turn alternation than was previously the case (Dember & Fowler, 1958; Lester, 1968), it is unfortunate that the number investigated is still relatively small. I hope that attempts will be made to improve the situation and that, particularly in studies of SAB, closer attention will be paid to the behavioral characteristics, capacities, and lifestyles of different organisms. There should be proportionately more investigations of SAB in invertebrates and turn alternation in vertebrates as well as studies of the mechanisms controlling both phenomena in all species.

Nevertheless, with the information we now have available, it is possible to identify some apparently general relationships across species. For those in which SAB has been established, comparisons of goal-arm stimuli experienced on successive trials seem more important than previous body turns. Although in most cases the particular sensory modalities involved have not yet been identified, as with rats (Dember & Fowler, 1958), a tendency to approach novel stimuli is the most likely motivational substrate for other species. Single or sequential alternation of body turns is primarily response-generated, but in some invertebrates tactile and visual stimulation can modify the phenomenon. When individuals of some species occasionally experience unusually close contact with a vertical surface, the resulting tactile stimulation might be sufficient to cancel out any effects of stored information about previous turns and thus determine the direction of a subsequent turn. The availability of both types of mechanism should provide some degree of flexibility for coping with the varying demands of different environments.

As a way of systematically varying behavior, SAB and turn alternation achieve similar results. In natural settings they also share some common adaptive functions but by different means. Because they are controlled by different mechanisms, SAB and turn alternation should in the future be treated as related but not identical phenomena.

# 4
# SAB, Reward, and Learning

CHARLES L. RICHMAN
WAKE FOREST UNIVERSITY

The primary purpose of this chapter is to document a meaningful relationship between two somewhat diverse areas of psychological research—discrimination learning and SAB. An attempt will be made to demonstrate that: (1) The cues responsible for SAB can be used to predict discrimination learning rates; and (2) Depending on the situation, SAB may interfere with or facilitate discrimination learning. A secondary aim of this chapter is to review and interpret the effects of food deprivation and food reinforcement on SAB, especially as they interact to affect the cues to which the rats are attending.

## *Intra-Maze and Extra-Maze Cues*

During the half century following Hunter's (1914) note regarding SAB and the two decades following Tolman, Ritchie, and Kalish's (1946) analysis of discrimination learning, it was assumed by most investigators that a subject's maze behavior was primarily determined by visual cues. This conclusion was, to say the least, ironic, since rats, the preferred test animal, were known to be nocturnal and ill-equipped to function adequately in daylight. The visual ("place") versus kinesthetic ("response") controversy in discrimination learning was somewhat resolved when Restle (1957) suggested that animals use all the relevant cues inside and outside a maze, with cue dominance depending upon the sensory capacity of the animal and the amount of relevant stimulation. A few years later we (Richman, Chappell et al., 1970) made a similar observation regarding SAB. Restle and Richman, however, were not the first to suggest that an animal's attention to various cues may vary as a function of several variables (Vincent, 1915).

There is little doubt that Douglas' (1966a) brilliantly conceived and executed series of studies had a powerful impact on our conception of SAB. Douglas demonstrated that a rat's tendency to alternate was a function of weak intra-maze cue, odor-trail avoidance, and a strong extra-maze cue, direction, and he appeared to deny any role of visual cues in SAB.

Although Douglas' (1966a) classic paper appeared to be the final say regarding SAB in rats, there remained several reports that suggested that visual cues may affect alternation in at least a modest way, e.g., O'Connell (1971), Pate and

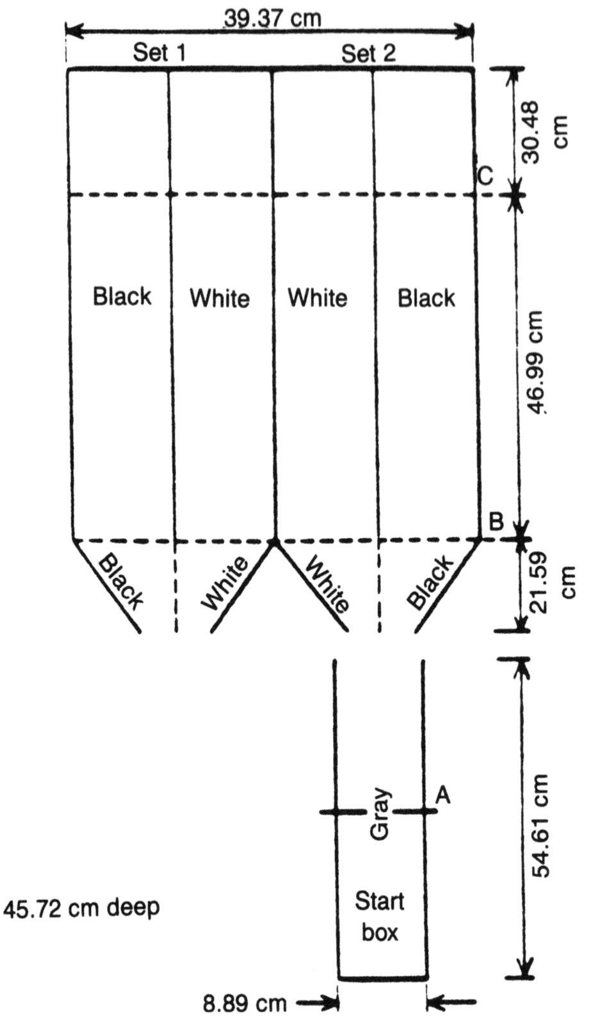


FIGURE 4.1. Plan view of parallel double-runway maze.

DeLoache (1970), Still and MacMillan (1969), and Thompson (1960). Furthermore, Dember (1956) and Kivy et al. (1956), employing a response-to-change procedure, found evidence suggesting that visual cues may have a strong effect on a rat's maze behavior.

Restle's (1957) compromise regarding place versus response learning was, in part, the impetus for our resolution of the cue controversy in SAB. In the Richman, Chappell et al. (1970) study, we argued that the kind of apparatus typically used in SAB research, T and Y mazes, predisposes rats to attend to directional cues and to ignore visual ones. We tested this assumption by assessing rats for visually versus directionally determined alternation in a T maze and a parallel double-runway maze, as depicted in Figure 4.1. (Note that the left runway in parallel maze #1 is painted black and the right one is painted white. However, the left and right runways in parallel maze #2 are painted white and black, respectively.) We found SAB to be a function of brightness cues in the parallel maze and of direction in the T maze. Thus, SAB was determined by the interaction of the type of maze and type of cue. The dominant cue in a T maze becomes nondominant (or less effective) in a parallel maze and vice versa. We interpreted the maze by cue effects in terms of attention theory. In a parallel maze, there is a high probability of rats' attending to visual cues and a low probability of their attending to directional cues; by contrast, in a T maze attention to directional cues is higher than to visual cues.

## *Cue Dominance in SAB: A Predictor of Discrimination Learning*

Based on the finding that rats' attention to specific cues could be inferred from SAB, Richman, Knoblock, and Coussens (1972) suggested that the presentation of cues as a function of the constraints of a maze could also predict discrimination learning rates. Our results supported this hypothesis. Rats learned a T-maze discrimination problem more rapidly to directional than to visual cues, whereas they learned a visual task in a parallel maze more rapidly than a directional one. Unfortunately, we used a between-subjects procedure so individual difference analyses across mazes and between relevant dimensions could not be assessed. It would be informative to perform individual-difference research, testing animals for SAB and discrimination learning, by varying the angles at the choice point as well as the brightness differences between the two goal arms. Here a subject's SAB and discrimination learning rates across several maze configurations and visual cue presentations could be documented. These results would provide additional evidence regarding the notion that cues that determine SAB are also predictive of discrimination learning rates. With sufficient data it might prove possible to determine precise psychophysical functions for each of the cue domains.

## *SAB and Discrimination Learning: Competing Response Patterns*

### *Early Research*

Historically, the behavior of animals during the initial trials of discrimination training were considered to be random (cf. Peterson, 1917). Dennis and Henneman (1932), however, argued that maze behavior was anything but random. They suggested that animals, prior to the onset of discrimination learning, engage in behaviors that lead to the elimination and/or the avoidance of predictable errors. For example, animals tested in elevated mazes initially demonstrate a position habit of returning to the same goal alley. Extinguishing a position habit must precede the learning of a discrimination problem (Jackson, 1941). Rodents tested in nonelevated mazes on brightness discrimination problems tend to be attracted to the darker goal arm and/or demonstrate an above-chance level of alternation behavior. These behavioral patterns must also be eliminated prior to the onset of discrimination learning (see, for example, MacGillivray & Stone, 1930). Position habits and SAB are obviously nonrandom, predictable behavior patterns that compete with efficient discrimination learning.

*Response Competition, Deprivation, and Reward*

If SAB competes with an animal's performance of a discrimination problem, then factors that decrease the probability of alternation should facilitate learning. Such an experiment was conducted by Richman, Gardner, Montgomery, and Benewicz (1970). In this study, we reported that T-maze discrimination-learning rates were directly related to body-weight loss and both were inversely related to SAB. Our severely food-deprived rats (76% ad lib weight) failed to alternate at an above-chance level during any of the initial three days of discrimination training and reached a 90% learning criterion in only 6.5 days. However, our low-drive animals (92.5% ad lib weight) took 9.5 days to reach criterion and showed an above-chance level of alternation during the first three training trials (10 trials per day). These data suggest that high drive facilitates learning by eliminating an animal's initial tendency to alternate.

In 1958, Estes suggested a similar relationship between levels of food deprivation and learning. Estes' (1958) model included the notion that with increasing deprivation levels there was a corresponding decrease in the importance of satiation cues associated with conditioned responses such as resting, inactivity, and lying down. These responses interfere with the response to be learned. A decline in performance following a reduction in the food-deprivation schedule was attributed to the tendency of these satiation responses to compete with the instrumental response. The major difference between our analysis and Estes' is that he considered the satiation responses as conditioned responses, whereas we consider SAB to be an unlearned response pattern.

## The Overtraining Reversal Effect (ORE) and SAB

The ORE has been documented literally hundreds of times in various laboratories and with numerous species (e.g., Clayton, 1963; Mackintosh, 1965; Richman & Coussens, 1970; Sperling, 1970). The defining characteristics of the ORE include training at least two groups of animals on a discrimination problem; for example, the subject must select the white but not the black goal arm in a T maze. One group of animals is overtrained, and a second group is trained to a lesser extent prior to reversal training—e.g., 100% criterion versus 80% criterion, respectively. Once the respective initial criteria are reached, the animals must select the previously nonrewarded goal arm. The ORE is evident when the overtrained animals are found to reach the reversal learning criterion in fewer trials than the nonovertrained ones. With rats, the ORE is almost always reported when difficult discrimination learning tasks are used, such as black-white T-maze discrimination tasks where spatial and odor cues are allowed to vary. With the exception of one study (Capaldi, 1963), the ORE has not been found in a simple, spatial discrimination task (e.g., Clayton, 1963; Lovejoy, 1966). This finding baffled many of us for years. After pursuing blind alleys too numerous to mention, we (Richman & Coussens, 1970) noted that Capaldi's (1963) training procedure differed from the norm in that Capaldi's nonovertrained rats received a mere 16 prereversal training trials. One of our hunches, which turned out to be correct, was that if rats are only given a few prereversal trials, the tendency to alternate remains dominant during the initial stages of reversal training, thereby delaying reversal training. That is, with only 16 original learning trials rats do not have sufficient training for their SAB tendency to drop out. One important service rendered through overtraining is to eliminate SAB. Without the interfering affects of SAB following overtraining, reversal training should occur quite rapidly, hence the ORE. The Richman and Coussens (1970) hypothesis was confirmed. Original learning trials were inversely related to SAB during reversal training, and as a consequence the ORE was found in a simple, spatial-discrimination task. More specifically, we found that undertrained rats exhibited above-chance levels of SAB during the first four days of reversal training and took approximately 2-½ more days to reach the reversal criterion than overtrained subjects. The overtrained animals showed no evidence of SAB during reversal learning.

The Richman and Coussens (1970) and Richman, Gardner et al. (1970) studies suggest that factors responsible for a reduction in SAB (e.g., high drive and/or overtraining) facilitate discrimination learning. Obviously, not all procedures that eliminate or reduce the probability of SAB facilitate learning. Hippocampal (Douglas, 1967), septal (Dalland, 1974), and amygdala (Rosen & Stein, 1968) lesions and certain psychopharmacological agents such as lithium chloride (Johnson, 1980), amphetamine and scopolamine (Kokkinidis & Anisman, 1976b) have been shown to reduce the probability of SAB, but do not have a consistent positive effect on discrimination learning. It would also be surprising if middle ear disease or shaking the animal back and forth between trials, which reduces the probability of SAB (Douglas, 1966b), would also facilitate discrimination learning. But perhaps not! For example, it is conceivable that learning may be enhanced in a T-maze brightness-discrimination task by surgical, pharmacological, or behavioral procedures that simply reduce T-maze SAB to chance without introducing a competing response pattern, such as perseveration. Here, the initial tendency to alternate is eliminated and the animal can "concentrate" on solving the brightness-discrimination problem. This finding would be consistent with the Richman, Knoblock, and Coussens (1972) parallel maze results, where directional alternation was minimized and a visual learning task was accomplished quite rapidly.

## *Discrimination Learning and SAB: Similar But Different*

So far, I have described situations in which the cues responsible for SAB predict discrimination learning rates and paradigms in which the elimination of SAB facilitates learning. This chapter would not be complete without a description of several studies that suggest a somewhat bizarre relationship between SAB and learning. First, however, I must point out some procedural differences between conducting SAB and simultaneous discrimination-learning studies. Most SAB research has been performed in a T or Y maze with sated, nonrewarded animals, whereas most often simultaneous discrimination-learning paradigms include rewarding a deprived (food, water, etc.) animal for making a specific goal-arm choice in a T or Y maze. In SAB we are concerned with the percentage of trials on which an animal avoids the previously entered goal arm; in discrimination learning we note the number of trials it takes for an animal consistently to reenter the same goal arm. As I mentioned earlier, it is not unusual for us to detect SAB during the initial discrimination-learning trials. Furthermore, I would not be surprised if SAB returned to a respectable level during the extinction of a learned response or if we simply free-fed the animals.

The studies to be reported here borrow from both the learning and SAB methodologies. The relevant question concerned the changes in the maze behavior of animals when they are food deprived and when they receive an equal amount of food for *either* choice-point turn. We did not consider this research to be a "what if" design for two reasons: (1) Although sparse, there were published reports tangentially related to this research; and (2) as the previous sections of this chapter suggest, we had good reason to expect that an intimate relationship existed between learning and SAB.

Unfortunately, published reports in this area are few. The empirical work by DeValois (1954) and the theoretical contributions of Montgomery (1953) had the greatest impact on our thinking. DeValois (1954) showed that high drive was associated with a decrease in path-selection variability. Montgomery (1953) proposed an inverse relationship between hunger level and extent of exploratory behavior (see also Zimbardo & Montgomery, 1957). The inference we extracted from those papers was clear: SAB should decrease as a function of an increase in drive. Subsequent reports only reinforced our suspicions. Fowler, Blond, and Dember (1959), using a forced-choice procedure, found that by the fifth test trial, animals who received large, relative to small, rewards for either goal-arm choice showed a significant reduction in alternation (see also Baker & Franken, 1968). And, finally, Fowler, Fowler, and Dember (1959) observed a change from SAB to perseverative responding as a function of intense electric shock presentations in the maze. Those studies suggested to us an inverse relationship between levels of drive/reward and SAB.

### *Study 1*

To assess this relationship in a free-choice situation, we (Richman, 1969) observed the cross-maze behavior of 360 male rats. The cross maze was used so we could assess the contribution of directional versus body-turn cues on SAB. Our design included two reward magnitude conditions (for either goal-arm entry rats either received 0 or 20 mg lab chow) and three levels of food deprivation (0, 24, or 48 hours).

Each subject was given five trials every third day for seven experimental days. A 15-second intertrial interval was used. The start stems were oriented S, N, S,

N, S on experimental days 1, 3, and 5 and N, S, N, S, N on experimental days 2, 4, and 6. Suppose the start stem was oriented S to N on trial 1 and the animal turned E (right); then, on trial 2, when the start stem was oriented N to S, if the rat turned W (right), his behavior was recorded as a direction alternation and a response repetition. If on trial 2 the animal turned E, a direction repetition and a response alternation were recorded. Under these conditions we did not know whether to attribute a consistent pattern of direction alternation to the animals' alternating to directional cues or to repeating their body turns. Therefore, on experimental day 7 all trials were conducted from the same start stem, e.g., N, N, N, N, N. (When I refer to the cross maze, I mean that the animal's start stem was reoriented 180° between trials, and T maze refers to a fixed-start stem.)

One other procedural point should be mentioned. Between trials a separate strip of heavy brown paper that lined the start stem was replaced with clean paper. For 120 animals the goal-arm paper was also replaced between trials with clean paper. For 120 other rats the goal-arm paper that the animal avoided on the previous trial was replaced with the paper from the goal arm he had just traversed; here direction and odor-trail avoidance were inconsistent. And finally, for the 120 remaining rats the paper remained in the same goal arm the animal had just traversed and a clean sheet of paper was placed in the nonselected goal alley between trials; thus, directional alternation and odor-trail avoidance were consistent for these subjects. I mention the paper-changing procedures to indicate that we were concerned with the possible confound of directional and odor cues in the interpretation of our results. However, odor cues were not sufficiently potent to alter our results and need not be discussed further.

The most interesting and relevant findings of the study were demonstrated on experimental day 7 (animals given all five trials from the same start stem); with the exception of one group, the animals continued to alternate direction at their previous high levels. However, for the 48-hour food-deprived-rewarded rats, direction alternation rates were found to be significantly *below* chance; that is, these animals were *response-repeating*. By contrast, the high-drive-nonrewarded animals continued to show *above*-chance level SAB on experimental day 7. Therefore, under conditions of high drive and reward (for making either choice-point turn), our animals showed above-chance-level *response repetition*. Alternation levels were equivalent between the reinforced and nonreinforced groups on experimental day 6 ($ps > 0.10$), and alternation levels were similar on experimental days 6 and 7 for the nonreinforced subjects ($ps > 0.10$). However, on experimental day 7 the reinforced group demonstrated statistically lower levels of SAB than the nonreinforced one; furthermore, the reinforced group's level of SAB was statistically greater on day 6 than day 7 ($ps < 0.05$). Table 4.1 shows the percentage of direction-alternation responses on experimental days 6 and 7 for the reinforced and nonreinforced 48-hour food-deprived animals.

Above-chance-level *response repetition* demonstrated by the high-drive, rewarded rats is analogous to at least three previously reported experimental outcomes. (1) Since reward is available for either choice-point turn, and the animals were not forced to make a specific turn (the only contingency for reinforcement

TABLE 4.1. Percent direction alternation on experimental day 6 (cross maze) and day 7 (T maze) for 48-hour food-deprived, reinforced and nonreinforced groups.

| | Day 6 (cross maze) | Day 7 (T maze) |
|---|---|---|
| % Reinforced | 76* | 39** |
| % Nonreinforced | 65* | 70* |

*Significantly above-chance-level direction alternation, $p < 0.05$.
**Significantly below-chance-level direction alternation, $p < 0.05$.

was to turn), the rats "unnecessarily" repeated the same turn response. Behaviorists might label this response pattern an example of superstitious behavior. Skinner (1948) stated that "the bird behaves as if there were a causal relation between its behavior and the presentation of food, although such a relation is lacking" (p. 171). Similarly, the rat that perseverates responses makes its turn as if this choice, and only this choice, leads to reinforcement, when, in fact, *either* turn would lead to food. Staddon and Simmelhag (1971) extended Skinner's (1948) interpretation of superstitious behavior by arguing that the superstitious response is isolated from other competing responses, stating that reinforcement selects "one response over others, or . . . (directs) a response which would probably predominate in any case" (p. 20). Thus, innate or preexperimentally learned behavioral tendencies could be influential in the occurrence of a particular superstitious behavior. (2) Response repetition is also analogous to the resultant perseverative responding induced by amphetamine. (3) These results may also be viewed in terms of the rat's adopting his preferred response in much the same fashion that an animal establishes contingent behavior when only one particular response is reinforced. The necessary condition for a repetitive preferred response occurrence are high drive and reward. The observation that rats under conditions of low drive maintained an above-chance level of SAB when equal amounts of food reinforcement were available on both sides of a T maze, but repeated their responses under high-drive-reinforced conditions, was explored further in a second unpublished study (Richman, 1970).

*Study 2*

One of the purposes of this study was to determine whether initial T-maze turn preferences were correlated with the side of the maze the animal chooses late in training. It is possible that under conditions of high drive, animals select their preferred body turn because it is initially more likely to be associated with reinforcement than the nonpreferred turn. Here, SAB is extinguished because during the initial reinforcement test trials the preferred turn has been reinforced on more trials than the nonpreferred turn. According to such a hypothesis, the animal tends to choose the initially preferred side late in testing. Let me quickly dispense with this hypothesis. We found no evidence to suggest that animals with an initial response bias would subsequently select (after 80 to 100 test trials) their preferred response.

Another reason we conducted Study 2 was to see whether the perseverative responding found in Study 1 would replicate. Here we extended our testing period to 100 trials. Thirty-three male rats were randomly assigned to one of two groups: 24 were placed in the 48-hour food-deprivation group and 9 were run in a four-hour food-deprivation group. All subjects were gentled for 5–10 minutes daily for three days and allowed to explore the start stem of the maze from 5 to 10 minutes for three additional days. The 48-hour food-deprived animals were run in the maze every third day with 24 hours of free feeding following each day's run. Food was removed from the four-hour food-deprived animals' cages four hours prior to each day's run. All subjects were maintained on ad lib water.

One nonreinforced trial per day for five days was conducted to determine each animal's turn preference. Following the subject's alley choice, he was returned to his cage; the maze was then cleaned with a damp sponge and the next animal was tested. The one-trial-per-day procedure was used to prevent normal alternation behavior from affecting the preference measure.

On experimental day 6 and for the next 10 days, each rat was given 11 trials per day (experimental days 6–15), which were initiated from the same start alley (e.g., S). The procedure for each trial was identical to the initial preference trials, except that reinforcement (20 mg Noyes pellets) was available at the end of both goal alleys. Subjects were given a maximum of two minutes to eat the food found in the goal box.

The results were markedly different for the 4- and 48-hour food-deprived subjects. Chi-square tests demonstrated that SAB was well above chance level ($p < 0.01$) during the 10 experimental days (days 6–15) for the 4-hour (low-drive) food-deprived group. Alternation was above chance level on each of these 10 days ($ps > 0.01$), with alternation rates ranging from 76 to 85% during the 10 test days. For the 48-hour (high-drive) food-deprived group, chi-square tests revealed significant alternation ($ps < 0.01$) during experiment days 6–9 (SAB = 78%, 81%, 78%, and 64%, respectively). Chance-level SAB was evident during experimental days 10 and 11, and below-chance-level SAB was demonstrated on experimental days 12–15 (SAB = 38%, 30%, 13%, and 7%, respectively). Correlated $t$ tests were conducted between the number of alternations on experimental days 6 and 15 for each deprivation group. These results revealed more frequent alternations on day 6 than on day 15 for the high-drive group ($t = 21.26$, $p < 0.001$) but no difference for the low-drive group ($t < 1.00$).

Thus, testing severely food-deprived animals in a T maze for 10 days ultimately resulted in a high level of response repetition and/or direction repetition. To assess the response versus direction repetition hypothesis we tested the animals on an 11th day (experimental day 17) in the cross maze and alternated the start stem orientations, e.g., N, S, N, S, N, etc. If our high-drive animals tended to repeat goal-arm entries (e.g., E, E, E, etc.), which would also indicate they were alternating body turns, we would assume the prior T-maze perseverative behaviors to be a function of directional cues. However, if they consistently alternated goal arms, thus repeating body turns, we would assume their T-maze behavior was a function of body-turn cues. The cross-maze results were clear: the animals repeated responses on approximately 9 out of 10 test trials (88%). In sum, Studies 1 and 2 strongly suggest that as a function of high drive, reward for either goal-arm choice and extended testing, a rat's maze behavior is determined by body-turn cues. The final level of response repetition found in Study 2 was similar to values found in simultaneous spatial discrimination investigations (e.g., Pubols, 1956).

The test results in the cross maze on the last test day for the 4-hour-rewarded animals were quite different from our high-drive rats. If you will recall, the low-drive animals consistently alternated in the T maze for 10 test days. We assumed that alternation was a function of directional cues. Thus, on the 11th day (the cross-maze-test day) they should also alternate on the basis of direction by repeating their body turns and avoiding their previously entered goal arm. This hypothesis was blatantly disconfirmed. In the cross maze our low-drive animals *repeated* goal-arm entries on more than three-quarters of the trials (76%). I can only assume that they were alternating responses during the last few T-maze-testing days.

Why the discrepancy between the results obtained in Study 1 and Study 2? In Study 1 our low-drive animal's alternation behavior was due to direction of movement; in Study 2 they alternated body turns.

Again we must borrow from the discrimination-learning literature. Seventy-five years ago, Vincent (1915) suggested that an animal's maze behavior is directed by exteroceptive cues early in training but proprioceptive cues later in training (see also Carr, 1917). Two discrimination-learning studies confirmed Vincent's (1915) thesis. Hicks (1964) and Mackintosh (1965, experiment 3) trained high-drive animals in a T maze with reward presented in only one goal arm (e.g., E and right goal box, and the start stem oriented S to N). They tested several groups of rats after varying the number of training trials. Once the original training criterion was reached, Hicks and Mackintosh reoriented the test trials start stems 180° (N to S). The only difference between the two studies was that Hicks' animals were not reinforced for choices during testing while Mackintosh's rats received 10 seconds of free feeding for either test trial choice. The results of both studies were identical in that extensive training increased the probability of response repetition; however with only a few initial training trials, their animals consistently selected the same goal arm. The former behavior pattern and the results from Study 2 (low-drive animals) are similar to those reported in single alternation paradigms where animals are rewarded in a T or Y maze for alternating responses.

In sum, the findings derived from Studies 1 and 2 suggest that under low-drive conditions rats alternate to directional cues early in testing, but with extensive, massed testing their behavior is controlled by proprioceptive[1] cues, resulting in

[1]Technically speaking, the proprioceptors are not only found in the muscles, tendons, and joints but they are also located in the labyrinth, giving information concerning the movements and position of the body in space (Sherrington, 1906). For our purposes, proprioception refers to the feedback given to the muscles, tendons, and joints by a particular body turn, with the limb acting as the stimulus to its own receptors.

TABLE 4.2. Predictions regarding the cues controlling high- and low-drive, rewarded rat's maze-selection patterns following extensive SAB testing.

| Overtraining in T maze | Subsequent cross-maze test | Overtraining in cross maze | Subsequent T maze test |
| --- | --- | --- | --- |

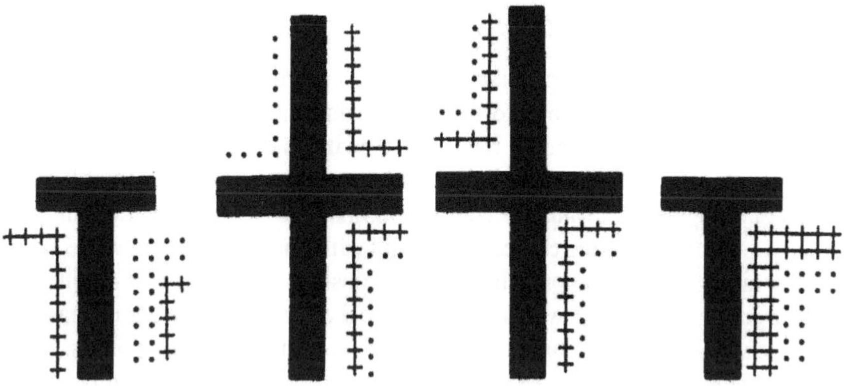

Notes: Trial 1 start stem oriented South to North.
High drive ·········
Low drive +++++++++

a body-turn alternation habit. Proprioceptive-determined behavior as a function of overtraining/overtesting results in response repetition for high-drive, rewarded animals and body-turn alternation for low-drive, rewarded animals.

The former hypotheses were adequately tested in Studies 1 and 2. The latter hypothesis can easily be assessed in the following experiment. Table 4.2 reports the predictions for rewarded high- and low-drive rats following overtraining in a cross or T maze.

Testing low drive rats for 25, 50, 75, 100, and 125 trials initially in a cross maze and subsequently in a T maze, we would expect that: (1) The rats who were administered only a few cross-maze trials would demonstrate SAB as a function of directional cues. Our rationale is quite simple: 25 test trials in the cross maze is an insufficient number of trials to establish a position habit. (2) When given 100 or more SAB cross-maze tests the animals' initial alternations are due to directional cues; however, later in testing a transition occurs that leads to a consistent body-turn bias (e.g., right turn and E, right and W, right and E, etc.). These animals should subsequently demonstrate *response repetition* in a T maze (e.g., right, right, right, etc.).

You will read in Chapter 8 that Douglas tested both rats and mice for a number of trials and found no decrease in their tendency to alternate. There are several differences to be noted between Douglas' and my procedures, any of which could affect the results: (1) He used distributed practice, and we used massed practice; (2) He did not reward his animals in the maze; we rewarded them for either choice; and, most importantly, (3) Douglas had no reason to, and therefore did not, test his animals' behavior with the start stem periodically reoriented 180°. Whether it is a discrimination-learning or SAB paradigm, I now consider it important to check the cues responsible for our animals' behavior by periodically reorienting the maze start stem 180°.

## *Theoretical Considerations*

This chapter was an attempt to show that the cues responsible for alternation behavior change as a function of drive, reward, and length of SAB testing. Unfortunately, there is no one theory that can adequately account for these findings.

Some years ago Nissen (1950) interpreted discrimination-learning problems in terms of an approach-avoidance conflict. In a similar vein, presenting animals with two equally tempting choices, as we did with our high-drive-rewarded rats, results in an approach-approach conflict. Consider the hungry rat facing the choice point with equal rewards in the two goal alleys. Here the animal has two positive goals, but he can only receive one of them at a particular point in time. To resolve the conflict and reduce the delay of reward, he consistently selects only one body turn. Heilizer (1978) has shown that subjects are slower to respond in an approach-approach situation than in approach, avoidance, or approach-avoidance situations, interpreting the slower response patterns as an index of conflict. Any delay, due to conflict, increases the time in which the rat delays eating. To eliminate this delay the subject avoids the conflict by selecting only one body turn. Predictions derived from this hypothesis include: (1) A more rapid transition from SAB to response perseveration as a function of an increase in drive with equal rewards present in the goal arms; (2) extinction of response repetition following the removal of rewards or a reduction of drive; and (3) under high-drive and reward conditions, greater vacillation at the choice point early in testing (during SAB) relative to the later stages of testing.

Hull's (1943) explanation of SAB was abandoned some 30 years ago, perhaps prematurely. According to Hull (1943), reactive inhibition is generated when a response occurs and then dissipates over time. Because reactive inhibition, an aversive drive, is derived from responding, the subject in an SAB situation alternates body turns on each trial. In Hull's theory inhibition accumulates with or without reinforcement. Furthermore, reactive inhibition should be strongest under low-drive and nonreinforced conditions (see, for example, Prokasy, 1960). In Study 2 (Richman, 1969), rats showed a tendency to alternate body turns when testing was massed and extensive and drive was low. These data appear consistent with Hull's (1943) reactive-inhibition model.

The Hullian model is not the only theory that can be used to describe our data. In Studies 1 and 2 we noted a behavior transition in our low-drive-rewarded animals. Early in testing our rats alternated to directional cues, but later they tended to alternate as a function of body-turn cues. Foraging models may also account for SAB. Most foraging models include a "win-shift" premise (Kamil & Sargent, 1981). That is, animals fail to return to a place where they previously found food. With low drive and equal rewards for either choice-point turn, foraging models predict a "win-shift" strategy. In foraging terms, the more time that animals spend sampling a patch, the less time they will spend exploiting it. If one considers the goal arm of a maze a patch, then low-drive animals should spend most of their time sampling the two "patches" and very little time exploiting them. Consistent with this prediction, our low-drive-rewarded animals alternated and only on occasions nibbled at the food pellets in the goal arms. I assume that patch sampling would decrease and exploitation would increase as a function of an increase in drive (see Study 2) and/or if the two goal alleys contained different amounts of food, a procedure used by Kacelnik (1979).

How does the foraging model deal with SAB when animals are sated and nonrewarded? Under these conditions we do not question the "shift," only what is "won" or "lost" prior to the shift (see Richman et al., 1987). In the typical SAB paradigm, a sated animal is placed in a T maze, visits on an alley on Trial 1, then visits the other alley on Trial 2. A "lose-shift" strategy reflects the animals' tendency to explore; when Trial 1 produces little stimulus change, sated animals follow an alternate route on Trial 2. Animals avoid familiar, nonstimulating environments and search for unfamiliar, more interesting surroundings. Furthermore, rats prefer running in different directions from trial to trial (e.g., Sherrick et al., 1979). Sherrick et al. (1979) suggested that the rats' tendency to run in a different direction on two successive trials facilitates their efficient exploration (sampling) of the environment (patches). Although the foraging model is provocative, I remain convinced that Dember's (1961) response-for-change model best describes a sated animal's tendency to alternate. A quote from our most recent review paper elaborates this position.

Whether animals note the alley previously visited via directional, odor, and/or brightness cues, it is evident that these cues permit them to identify the two alleys. Once animals distinguish and store information pertaining to a visited and an unvisited alley, they subsequently explore the unvisited one. In the response-to-change experiment (Dember, 1956), animals explore the novel (changed) alley; in an SAB paradigm they enter the alley that provides them an opportunity to discover a novel stimulus: they "respond-for-change" (Richman et al., 1987, p. 385).

# 5
# Spontaneous Alternation Behavior and the Brain

ROBERT J. DOUGLAS
THE UNIVERSITY OF WASHINGTON

One strategy for writing this chapter would have been to prepare an immense list of all studies linking spontaneous alternation behavior to brain lesions or drugs. Our minds thoroughly boggled, we could then have identified a few missing items in the list and agreed that further research was necessary. I believe, however, that even though there are many studies crying out to be done, it is possible to make considerable sense out of the staggering amount of evidence already in existence. The results of hundreds of alternation experiments do tell a story—two quite different stories, in fact. And these stories reveal more than why a rat turns right or left in a T maze. The alternating rat is trying to tell us how the brain works, and my intent in writing this chapter is to convey to you what the alternating rat tells me.

The most basic way of studying brain function is to destroy a given region or structure and see what happens. Such studies have shown that while SAB is definitely not affected by brain damage in general, there are in fact many brain structures that appear to be involved in alternation. This can be bewildering if one is unaware of the anatomical relationships between these seemingly different structures, which can be assembled into systems on the basis of interconnections. As we shall see, it is possible to align each alternation-related structure with either of two systems, one of which is centered on the hippocampus and the other on the vestibular system. The problem then becomes one of trying to determine what each of these sets contributes to SAB and what each structure contributes to its set. This procedure is facilitated by considering exactly what is involved when a rat alternates in a T maze. First, the animal must be able to apportion its perceptual world into two parts, recognize the difference between them, and remember which one has been most recently visited. Second, the animal must have some motive for avoiding revisits. Call it curiosity. But another motive is also often operative in the T maze, and "fear" is a good name for it.

## *Fear and Perseveration*

It has been known for almost half a century that fearfulness antagonizes alternation (e.g., Heathers, 1940). Rats often perseverate when placed in a two-choice situation involving danger. They do not alternate when forced to leap into space

(Jackson, 1941) or (in my lab) when tested in vertical or water-filled T mazes. In one study (Douglas, Kowal, & Clark, 1968) ungentled rats were formally rated for emotionality on the basis of behavior. The animals in the most-emotional category averaged only 10% alternation—or very strong perseveration. It makes intuitive sense that a terrified animal should preferentially visit the most *familiar* locale rather than the most novel. Thus, any given brain lesion might spuriously knock out alternation by producing emotionality. Olfactory bulb lesions are a good case in point.

### *Olfactory Bulb Lesions*

Large bilateral lesions of the olfactory bulb result in a significant drop in SAB, but they also tend to produce "irritability" in some rats and extreme emotionality in others (Douglas, Isaacson, & Moss, 1969). In that study the lesions were found to reduce alternation in proportion to the degree of emotionality, and as the emotionality waned with the passage of time (weeks), alternation increased to a normal rate in most cases. Similar results were found by Richman, Gulkin, & Knoblock (1972). In a later unpublished study I found that the loss of olfaction per se had no detectable effect on either emotionality or alternation. Irrigation of the nasal mucosa with a zinc sulfate solution abolishes olfaction without producing direct brain damage (Alberts & Friedman, 1972). This procedure was employed bilaterally in eight rats, with each being given six alternation tests within the following two days. None of the animals became hyperemotional, and the group averaged 83.3% alternation.

Conceivably, a lesion could abolish SAB both directly and, through increasing emotionality, indirectly. This is probably the case with septal-lesioned rats, which may strongly perseverate when displaying "septal rage" but still fail to alternate many weeks later when they have calmed down (Douglas & Raphelson, 1966). Drugs, too, can produce terror as an unintended side effect, and it is unfortunate that in the following analysis we cannot always be sure that emotionality was not involved. With that caveat, we are ready to launch an investigation into brain regions that are directly involved in alternation, and the hippocampal system is the best place to begin: Virtually every theory of hippocampal functioning predicts that hippocampal lesions ought to eliminate SAB.

## *Alternation and the Hippocampal System*

### *Hippocampal System Lesions*

Table 5.1 displays some of the data I have gathered over more than two decades of investigation. It can be seen that massive bilateral lesions of the hippocampus totally abolish alternation. Despite the large number of animals and the numerous tests given to each one, the alternation rate is not reliably different from chance. This is true not only of the group as a whole, but also of individuals, not one of which reliably deviates from chance. Many other studies have reported basically similar results (e.g., Dalland, 1970; Douglas & Isaacson, 1964; Ellen

TABLE 5.1. Spontaneous alternation rates (raw or unadjusted for chance) in 354 rats given standard 2-trial tests in T mazes of similar size and construction. Eleven animals received 10–15 tests each, but the others had a minimum of 20.

| Description of lesion | N | % Alternation |
|---|---|---|
| 1. Unlesioned, well-gentled rats | 149 | 85.8** |
| 2. Massive bilateral hippocampal | 58 | 44.8 |
| 3. Same, but in two stages | 10 | 47.3 |
| 4. Complete fornix transection | 4 | 47.1 |
| 5. Massive bilateral septal lesions | 6 | 43.8 |
| 6. Small electrolytic hippocampal | | |
| A. Week 1 postoperative | 12 | 49.7 |
| B. Week 2 postoperative | 12 | 62.2* |
| C. Weeks 3-9 postoperative | 12 | 72.7** |
| D. 1-3 months postoperative | 6 | 76.3* |
| 7. Partial (mainly lateral) fornix | | |
| A. Week 1 postoperative | 4 | 54.2 |
| B. Week 2 postoperative | 4 | 71.9 |
| C. Weeks 3-5 postoperative | 4 | 80.3* |
| 8. Massive unilateral hippocampal, all | 10 | 83.6** |
| A. Left hippocampal lesion | 5 | 86.4** |
| B. Right hippocampal lesion | 5 | 80.9** |
| 9. Neocortex, nonfrontal, bilateral | 52 | 83.2** |
| 10. Neocortex, unilateral | 14 | 83.7** |
| 11. Neocortex, dorsolateral, frontal | 7 | 86.3** |
| 12. Bilateral cingulate gyrus | 8 | 80.1** |
| 13. Bilateral amygdala | 12 | 88.2** |
| 14. Large unilateral caudate nucleus | 2 | 00.0 |

*Significantly above chance at the 5% level.
**1% level or better.

& DeLoache, 1968; Isseroff, 1979; Kirkby, Stein, Kimble, & Kimble, 1967; Lash, 1964; Laughlin, Finger, & Bell, 1984; Means, Leander, & Isaacson, 1971; Roberts et al., 1962; Stevens, 1973; Stevens & Cowey, 1973; etc.).

In two of these studies (by the Kirkby group and Stevens) it was suggested that the lesioned animals might alternate normally if given extra time for memory to "sink in." That is, the hippocampectomized rats were reported to alternate more (or perseverate less) if detained in the first-chosen alley for 50 minutes instead of 50 seconds. In neither study, however, did SAB at the long interval differ reliably from chance. Moreover, variations in detention time (within reasonable limits) have no detectable effect on alternation in normal rats. Finally, Denis Mitchell and I tried to replicate these results (unpublished data) and found a 50-minute detention in a cramped maze alley to be extremely aversive to the rats, who tried frantically to get out and produced astonishing amounts of urine and feces.

Assuming that emotionally distraught rats avoid odors suggestive of fear, their avoidance of the first-chosen alley would not represent true SAB and might be exhibited even by totally amnesic animals.

*Possible Recovery of Function*

The data in Table 5.1 explain several apparently discrepant results in the literature. That is, small lesions do not abolish alternation if the animal is given sufficient time to recover from surgery. The evidence suggests that three weeks is a conservative estimate for "sufficient time." The phenomenon termed "diaschesis" is well known to neurosurgeons. For example, a patient receiving only slight damage to Broca's area often awakens the next morning unable to speak. Speech then recovers gradually over the following days or weeks. Apparently something similar happens to rats following small (usually electrolytic) hippocampal lesions or partial transections of the fornix. Note, however, that even after several weeks there appears to be a partial loss of SAB, or a reduction of perhaps 10%. Kimble (1976) found recovery of alternation to occur in his small-lesioned rats after only two weeks, with changes in locomotor activity due to the lesions also normalizing at that time. Kimble and others have speculated that to some extent this recovery might be due to reinnervation of remaining tissue, but this remains a controversial point. In contrast to this recovery after small lesions is the apparent permanence of the effects of large (67–90%) lesions. Even when recovery is maximized through the use of two-stage lesions, with weeks of recovery between stages, these almost-complete lesions appear to produce a permanent deficit.

On the other hand, some studies have shown what superficially appears to be a recovery of alternation after very large hippocampal lesions (Isseroff, 1979; Isseroff & Isseroff, 1978). In these studies an apparently complete recovery of alternation occurred when the intertrial interval was zero but not when it was 10 seconds. This was interpreted as supporting the idea that the hippocampal contribution to alternation must be mnemonic, since at the zero interval the rats could employ the nonhippocampal mechanism of short-term memory. On the other hand, I have personally tested hippocampally lesioned rats with a zero intertrial interval and have found no evidence of increased alternation. A problem, however, is that even though the intertrial interval is "officially" zero, the actual time taken prior to a decision might easily exceed short-term memory limits.

In contrast to my negative results mentioned earlier, I have also gathered some unpublished evidence that appears to support the Isseroffs. In the process of testing mice in an automatic maze (see my other chapter in this book) I found that scopolamine was not totally abolishing alternation like it was supposed to. In analyzing the latency data I found that the mice alternated at a rate above 70% when the time from leaving one alley exit to entering the next was less than three seconds and at a rate slightly below 50% when the time was three seconds or more. These observations have potential importance but should be firmly established prior to further speculation.

*The Hippocampus as a System*

The evidence as a whole indicates that one must view the hippocampus as a system rather than as merely one structure. This system includes entorhinal cortex, subiculum, hippocampus proper(CA1-4), dentate gyrus, fornix-fimbria, and septal region. Table 5.1 shows that SAB can also be abolished by massive septal lesions or complete fornix transection. This is not surprising, since with few exceptions the effects of septal and hippocampal lesions on a wide range of behavior are virtually identical (see Gray & McNaughton, 1983). Alternation is also abolished by killing 85% of the neurons of the dentate gyrus by X-irradiation in infancy (Bayer, Brunner, Hine, & Altman, 1973). Destruction of region CA3 by 3-acetyl pyridine also knocks out SAB (Blampied & Wilby, 1975), as do large bilateral lesions of entorhinal cortex (Gibbs, Yu, & Cotman, 1987; Loesche & Steward, 1977; Steward, Loesche, & Horton, 1977). In the former study SAB was still at a chance level six months after surgery.

The hippocampal system thus appears to function as an interdependent series of substructures, each of which is critical for the functioning of the whole. This principle also appears to apply to the human hippocampus, since the well-known loss of recent memory has been shown to occur with a lesion restricted to, but almost totally destroying, sector CA1 (Zola-Morgan, Squire, & Amalal, 1986). In summary: (1) SAB is reduced or abolished after large bilateral lesions of any subdivision of the hippocampal system, but (2) It recovers to varying extents when the lesions are small or incomplete (Johnson, Olton, Gage, & Jenko, 1977; Kimble, 1976; Myhrer, 1975b; Ramirez & Stein, 1984; Will, Duluzarche, & Kelche, 1983; etc.). Although further research is needed to establish these points, the medial septum and medial fornix seem to be more critical than the lateral septum or fornix, and the precommissural fornix appears to be more critical than the postcommissural (Clody & Carlton, 1969; Johnson, et al., 1977; McNaughton & Feldon, 1980; Myhrer, 1975b).

*Lesions of Cortex or Other Limbic Regions*

Table 5.1 also reveals that SAB is not affected by lesions in general and that nothing akin to Lashley's mass action is operative. Many of the 59 rats with bilateral neocortical lesions had more extensive brain damage than some of the animals with hippocampal system lesions, and yet they alternated at very high rates. It can also be seen in Table 5.1 that SAB is not affected by limbic system lesions in general. The eight rats with cingulate gyrus damage alternated at a rate of about 80% whereas the dozen amygdalectomized animals averaged an astonishing 88%. The latter finding is supported by the study of Myhrer (1975a). He bilaterally severed the stria terminalis, a major output pathway of the amygdala, and found this lesion to have no effect on alternation. Thus, as far as SAB is concerned the limbic system certainly does not function as a whole.

It is, however, equally important to emphasize what the results do *not* show. Although 59 rats with bilateral neocortical lesions sounds like a lot, in fact most had lesions of the patch of cortex overlying the hippocampus. I strongly suspect that complete decortication would abolish alternation and that there are cortical regions that might be critical but have not yet been studied. It is worthy of note that none of the 59 rats with bilateral cortical lesions or the 14 with unilateral had extensive lesions of what would be the parietal lobe in a primate. It therefore remains possible that cortical lesions might produce "neglect" in rats (and a lack of alternation) if properly located. Table 5.1 also shows that the seven rats with bilateral lesions of frontal cortex averaged about 86% alternation, which superficially appears to conflict with the findings of Divac, Wikmark, & Gade (1975). The latter study, however, found alternation to be significantly reduced in rats with lesions of midline frontal cortex, whereas my rats all had lesions of dorsolateral frontal cortex. In rats the midline frontal cortex is a target of the ascending dopaminergic system and receives input from hippocampal sector CA1 (Swanson, 1981) and the subicular/entorhinal regions (Van Hoesen, 1985). As will be shown later, the bulk of the evidence argues against a frontal involvement in SAB except for this special midline region.

Finally, my cingulate gyrus lesions were relatively small, shallow, and about a third of the way back from the genu of the corpus callosum. It therefore remains possible that larger, deeper, more extensive, or differently placed cingulate lesions might affect SAB. For example, the midline frontal region mentioned earlier could be called the anterior cingulate gyrus. To my knowledge, there have been no alternation studies involving the strongly hippocampus-related retrosplenial region. The cingulate gyrus in general, however, does not appear to be very promising, since cingulate lesions have been found to produce a number of behavioral effects that are opposite to those seen after hippocampal lesions (Woodruff, Baisden, & Douglas, 1981).

*Hippocampal Function in Man and Beast*

With only slight recourse to sophistry it is possible to show that the hippocampus does basically the same thing (or things) in rats and in people. The inhibition theory (Douglas, 1972, 1975) applies equally well to man and beast, as do the "working memory" (Olton, 1983) and "spatial memory" ideas (O'Keefe & Nadel, 1979). Table 5.1 suggests, however, that there is one major difference. In humans the left hippocampus appears to be involved in verbal memory and the right hippocampus in pictorial/spatial memory (Milner, 1985; Smith, 1987). This appears to be a manifestation of the high degree of lateralization in the human brain, with verbal/sequential functions dominating the left hemisphere and spatial/simultaneous/holistic processes the right. Rodents do not, of course, possess verbal functions, but it remains conceivable that the left hippocampus might be especially concerned with temporal processes and the right with spatial, in which case one might expect right hippocampal lesions to knock out SAB. In fact, however, rats with left or right hippocampal damage in Table 5.1 are virtually identical in their alternation rates.

It should be noted, however, that those animals were given preliminary pretests immediately after surgery and that they all had very strong turning and/or position biases that persisted for several days. Specifically, all animals turned ipsilaterally or towards the lesion. At the time, I attributed this to a temporary inactivation of the caudate nucleus due to the necessary rupture of the lateral ventricle during surgery. Saporta & Greene (1974) discovered, however, that a directional orientation bias may remain long after surgery. That is, the rat with a unilateral hippocampal lesion preferentially orients to ipsilateral stimuli (and/or ignores contralateral) if simultaneously presented with equal stimulation on either side. This is a rodent version of the phenomenon called "extinction" that remains after a human patient recovers from contralateral sensory neglect. I did not test my rats for preferential attention, but they displayed no turning/place biases after those first few days and behaved in every observable respect like normal rats in the T maze.

Thus, present evidence overwhelmingly points to the hippocampal system as being crucial to SAB. Remaining to be explored, however, is a question of definition: Where do we stop in our inclusion of structures in the hippocampal system? Do we include the classical "Papez circuit" (Papez, 1937) or the various inputs and outputs emphasized in recent years? This problem will be addressed in more detail later. At this point we should perhaps stand back and look at the larger problem of hippocampus-related behavior. In animals, bilateral hippocampal lesions produce significant changes on numerous tasks or problems (see Douglas, 1967, 1972, 1975; Isaacson, 1974; Kimble, 1968, 1975; O'Keefe & Nadel, 1979; Olton, 1983; Olton, Becker, & Handelmann, 1979, etc.). Nearly all of these effects can be predicted by a model in which the hippocampus is postulated to mediate Pavlovian internal inhibition. The rule of thumb is that if inhibition facilitates performance on a task, then rats with hippocampal lesions will do poorly. If the task does not involve inhibition, the lesioned animals will be roughly normal. If inhibition interferes with learning or performance, then hippocampal damage will actually improve performance (Douglas, 1967, 1972, 1975).

According to the inhibition theory, SAB is merely a special case of "extinction of the orienting reflex" or one type of habituation. In any event, it follows that if a given brain lesion produces several items from the "hippocampal syndrome" list, then one would expect it to abolish SAB. Conversely, if a lesion *appears* to affect SAB but does not produce other hippocampal syndrome effects, then one might suspect the alternation effect to be spurious. It should be pointed out that SAB is not merely one of many hippocampus-related behaviors. Instead, it is one of the handful of behaviors that are hippocampus-dependent. The latter refers to behavior that is not merely reduced, but is actually abolished by hippocampal damage. In addition to SAB, hippocampus-dependent tasks probably include the radial maze, shocked T maze, and DRL (delayed) lever-press.

In human beings the best example of a hippocampus-dependent process is recent memory. People with bilateral hippocampal lesions possess normal short-term memory and remember many things from the remote past, but they are simply unable to consciously remember their ongoing experiences a minute or so later (Milner, 1985). They are, however, able to learn a number of tasks that they perform correctly despite being unable to consciously remember what they have done. These include two-stimulus discrimination, learning a simplified maze, mirror tracing, and learning to identify incomplete pictures (Milner, 1985; Milner, Corkin, & Teuber, 1968; Sidman, Stoddard, & Mohr, 1968).

Many seemingly different theories of hippocampal functioning do not actually compete with each other in the sense that if one of them is correct, then the others must be wrong. The inhibition theory, for example, refers to a mechanism that underlies performance on many different tasks whereas many other theories primarily identify specific tasks or situations in which the hippocampus appears to be important. As an example, Weiskrantz and Warrington (1975) present compelling evidence that their human amnesic subjects are unable to overcome interference and that they can perform almost normally on memory tests when interference is minimized. One could then state that the hippocampus is involved in overcoming interference. The inhibition theory can accommodate those data by postulating that interference is normally overcome via inhibition of competing memories and/or pathways.

In summary, a brain structure or pathway could be said to be part of the hippocampal system if its destruction produces the hippocampal syndrome. We can then examine whether the alternation system is coextensive with the hippocampal system and/or with the recent-memory system. We will begin with a consideration of structures or pathways that are anatomically and behaviorally related to the hippocampus. We shall begin with input pathways that are associated with certain transmitters and thereby sensitive to drugs affecting those transmitter systems.

## Hippocampal Inputs and SAB

### *The Cholinergic System*

The effects of drugs on spontaneous alternation are best understood in terms of hippocampal inputs, some of which are shown in Figure 5.1.

The best-studied hippocampal afferent pathway is the cholinergic input from the medial septum-diagonal band complex, a pathway that produces smooth "sinusoidal" theta waves (see Stumpf, 1965). Drugs that block muscarinic cholinergic synapses abolish this smooth low-frequency theta, while drugs that stimulate these synapses enhance theta waves. There is, however, a second type of hippocampal rhythm that could also be called "theta" (see Vanderwolf, Kramis, Gillespie, & Bland, 1975). This other type of theta wave is of a higher frequency, is not affected by anticholinergic drugs, and tends to be associated with locomotor behavior. The latter suggests that the hippocampus might be concerned with movement-generated feedback relevant to navigation. Cholinergic theta, on the other hand, has no obvious behavioral correlates.

Anticholinergic drugs such as scopolamine can completely abolish SAB even at relatively low doses (e.g., 0.3 mg/kg in rats: Douglas & Truncer, 1976). Studies reporting the abolition or great reduction of alternation as a consequence of

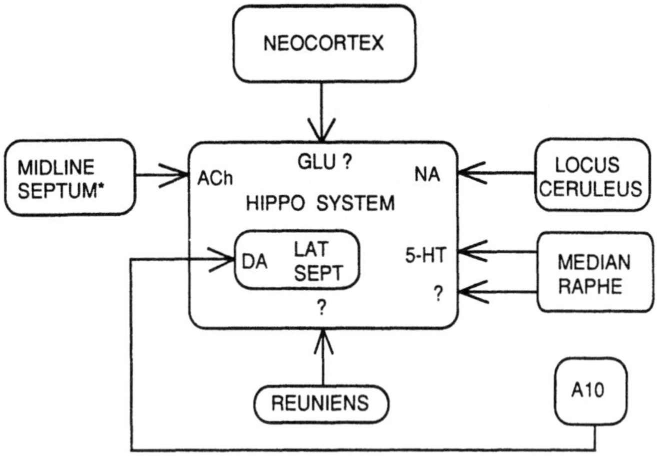


FIGURE 5.1. Major inputs to the hippocampal system. GLU = glutamic acid; ACh = acetyl choline; NA = nor-adrenaline; DA = dopamine; 5-HT = serotonin.

anticholinergic drug injection include Anisman (1975), Anisman and Kokkinidis (1975), Douglas and Isaacson (1966), Drew, Miller, and Baugh (1973), Hughes and Greig (1975), Kokkinidis and Anisman (1976b), Leaton and Utell (1970), McNaughton and Feldon (1980), Meyers and Domino (1964), Parkes (1965), Sarter, Bodewitz, and Stephens (1988), and Tobe, Egawa, and Nagai (1983). In human research anticholinergic drugs produce the same loss of recent memory seen in people with verified hippocampal damage (Drachman, 1977; Hammond, Meador, Aung-Din, & Wilder, 1987; Ketchum, Sidell, Crowell, Aghajanian, & Hayes, 1973; Meador, Loring, Adams, Patel, & Davis, 1987).

Drugs that stimulate cholinergic synapses have been found to enhance SAB (Baisden et al., 1972; Douglas, 1975; Egger, Livesey, & Dawson, 1973; Kokkinidis & Anisman, 1976b; Squire, 1969). A major problem with drug enhancement of SAB is that the rate is normally so high that it is almost impossible to statistically demonstrate an increase. Thus, in the cases cited here the investigators took advantage of situations in which the undrugged alternation rate was considerably reduced (e.g., long intertrial interval, "stunted development," very early age, poorly alternating species, etc.). In a lifespan developmental study of hippocampus-dependent behavior in mouse and man, SAB was found to drop to a near-chance level in aged mice and to be restored to almost its youthful level by the cholinergic stimulant physostigmine (Douglas, Douglas, & Treneer, 1979).

The cholinergic system is, however, not entirely dedicated to the hippocampus. There are numerous cholinergic interneurons in the corpus striatum, for example, and a large continuous cluster of long-axon cholinergic neurons termed "nucleus basalis" is located generally ventral to the pallidum (see Schwaber, Rogers, Satoh, & Fibiger, 1987). In this chapter, the term "nucleus basalis" will be taken to include all long-axon cholinergic forebrain neurons, and most of these innervate neocortex rather than the hippocampus. Thus, the drug evidence does not by itself prove that the cholinergic input to the hippocampus is crucial. A number of studies have shown, however, that SAB can be abolished by small lesions confined to the medial septal region (Clody & Carlton, 1969; Johnson et al., 1977; McNaughton & Feldon, 1980; Thomas, 1979). Hess and Blozovski (1987) found SAB to be reduced or abolished by direct injections of anticholinergic drugs into either the dentate gyrus, hippocampus proper, or entorhinal cortex. The evidence thus strongly suggests that anticholinergic drugs abolish SAB by blocking a crucial cholinergic input to the hippocampus rather than cholinergic pathways from basalis to neocortex. It is, however, a distinct possibility that the cholinergic innervation of neocortex mediates a function in which the hippocampus participates.

*The Raphe Input*

Another potentially important input to hippocampus is from the median raphe nucleus (centralis superior). Although the raphe nuclei are notable for their serotonergic neurons, they also contain many neurons that do not secrete this transmitter (Kohler & Steinbusch, 1982). Nonselective electrolytic lesions of the median raphe (but *not* of other raphe nuclei) produce the hippocampal syndrome, including the abolition of spontaneous alternation (see, e.g., Geyer, Puerto, Menkes, Segal, & Mandell, 1976). Recent evidence indicates, however, that the critical raphe neurons are not serotonergic (Asin & Fibiger, 1984). That is, alternation is abolished by procedures that unselectively kill median raphe neurons in general (electrolytic lesions, ibotenic acid) but *not* by those that selectively destroy serotonergic neurons in particular (the neurotoxin 5,7-DHT). Thus, it would appear that a pathway from the median raphe to the hippocampus is crucial for alternation but that serotonin is not the transmitter in this pathway. It remains possible, of course, that the crucial factor is a combination of the serotonergic and the unknown-transmitter pathway acting together.

*Catecholaminergic Inputs*

The destruction of dopamine neurons by 6-OHD injections in infancy has been found to have no effect on SAB in rats over 21 days of age tested with home-cage sawdust in the maze (Freeser & Raskin, 1982), though the treatment may have delayed development for a few days and/or enhanced emotional reactivity. Galey, Durkin, Sifakis, Kempf, and Jaffard (1985) specifically destroyed the dopamine pathway from nucleus A10 to the lateral septal region and found SAB to be increased by the lesion. They suggested that the septal dopamine input might inhibit cholinergic neurons. Taghzouti, Simon, and LeMoal (1986) used the same technique of 6-OHD injections into the lateral septum but found this procedure to reduce SAB to about 58% in food-satiated rats. Strangely, SAB bounced back to normal when the animals were food-deprived. Thus, there is no solid evidence suggesting that the dopamine input to the hippocampal system plays an indispensable role in SAB.

Mason, Wood, and Angel (1983) employed a technique in which neonatal 6-OHD injections produced a 97% reduction of nor-adrenaline in hippocampus and neocortex without affecting brain levels of dopamine. SAB rates were virtually identical in the lesion and control groups. On the other hand, Kokkinidis (Chapter 6) found a reduction in three-alley alternation after 6-OHD lesions of locus coeruleus. The lesioned animals were, however, not reliably different from the controls after the first two test days. Since MacNaughton and Mason (1980) cite numerous cases in which the behavioral effects of dorsal nor-adrenergic tract lesions differ from those of hippocampal damage, one might tentatively conclude that this pathway is of less than overwhelming importance to hippocampal functioning.

When drugs affecting the catecholaminergic systems are used systemically they also generally fail to affect SAB when employed at moderate dosage. Thus, the stimulant amphetamine has no effect at all on SAB at doses of 1 or 2 mg/kg, which are more than sufficient for the production of observable changes in other behaviors (Adkins, Packwood, & Marshall, 1969). The Adkins group found SAB to be affected only at the very high doses of 3.3 or 4 mg/kg, which produce compulsive perseverative behaviors. Many studies have found a variety of catecholamine stimulants or blockers (e.g., neuroleptics) to have no effects on SAB when employed at what I consider to be reasonable doses (Douglas, 1975; Grandjean & Battig, 1962; Parkes, 1965; Sinha, Francks, & Broadhurst, 1958; Swonger & Rech, 1972; Zahner, Battig, & Grandjean, 1961). These drugs also fail to produce anything like the recent-memory loss seen in humans after anticholinergic drug treatment or hippocampal damage (e.g., Liljequist, Linnoila, & Mattila, 1978). As we shall see, the corpus striatum appears to be involved in spatial orientation and SAB, and the dopamine input to the striatum is crucial to its functioning. Many of the behavioral effects of destroying or blocking the striatal dopamine inputs require, however, that the pathway be almost completely knocked out.

In summary, evidence linking the hippocampal cholinergic input to SAB is overwhelming. Strong evidence also suggests that a hippocampal input from the median raphe nucleus (but probably not serotonergic) is critically involved in SAB. No compelling evidence links SAB with catecholaminergic inputs to the hippocampal system. Strangely, the pathway from the thalamic reticular formation to the hippocampus (nucleus reuniens, see Herkenham, 1978) has yet to be behaviorally investigated. The profuse projections from neocortex to the hippocampal system are almost certainly of great functional significance, but the selective and complete severing of these pathways would be technically very difficult to accomplish.

## *Hippocampal Outputs*

Although many investigators appear to believe that the fornix is the output pathway from the hippocampus, the subiculum and entorhinal cortex project powerfully in the opposite direction to innervate various neocortical regions (Van Hoesen, 1985). Neocortical lesions do not produce the hippocampal syndrome, and neocortex could not be placed within a hippocampal system if the latter were to have any meaning. We will, however, address the problem of whether any particular set of "downward" connections appears to mediate alternation and/or known hippocampus-related behavior. A number of these pathways have been presented in classical papers by Nauta (1958, 1960). One powerful descending projection travels via the postcommissural fornix to the mammillary nuclei. A second set of pathways usually involves a septal synapse and descends (with many variations on the theme) either via the habenula and associated habenulo-interpeduncular tract on the one hand, or the medial forebrain bundle on the other. A simplified version is shown in Figure 5.2.

Interestingly, some elements of both pathways end up in what Nauta calls the "limbic midbrain region". There is no a priori reason to suspect that one of these two major routes should be more important than the other except for two historical accidents. The first was the highly influential speculative paper by Papez (1937), which focused attention (for all the wrong reasons) on the "Papez circuit." The second was that people knew about the Korsakoff syndrome at a time when nucleus basalis and the cholinergic system were virtually unknown. This syndrome, associated with severe alcoholism and malnutrition, includes memory defects similar to those observed in people with bilateral hippocampal damage (see Talland, 1968). The Korsakoff syndrome has traditionally been explained in terms of the high frequency with which lesions are observed in the mammillary bodies at autopsy (e.g., Brion, 1969). Recently, however, many people have apparently bought the theory of Victor, Adams, and Collins (1971): that lesions of the dorsomedial thalamic nucleus are critical to the memory disorder. This is strange because the dorsomedial (or mediodorsal) nucleus is primarily associated with prefrontal cortex and the amygdala, and lesions of these areas do not produce anything like the amnesia seen in the Korsakoff syndrome. For years arguments raged over whether the Korsakoff syndrome was due to (1) mammillary body lesions, (2) dorosmedial damage, or (3) both. As fate would have it, the correct answer was (4) none of these. Recent evidence shows that the Korsakoff syndrome is, like Alzheimer-type senility, associated with extensive cell loss in nucleus basalis, including the cholinergic pathway to the hippocampal system (Arendt, Bigl, Arendt, & Tennstedt, 1983).

### *The Postcommissural Fornix*

Meanwhile, however, the ideas about Korsakoff-related brain damage led many investigators to study the effects of mammillary body and/or dorsomedial thalamic lesions on the hippocampal syndrome. One important output of the mammillary nuclei is the ascending Papez circuit projection via the mammillothalamic tract to the anterior thalamic nuclei, which then project to the cingulate gyrus. Another is the descending mammillotegmental tract, which innervates such brain stem regions as the limbic midbrain.

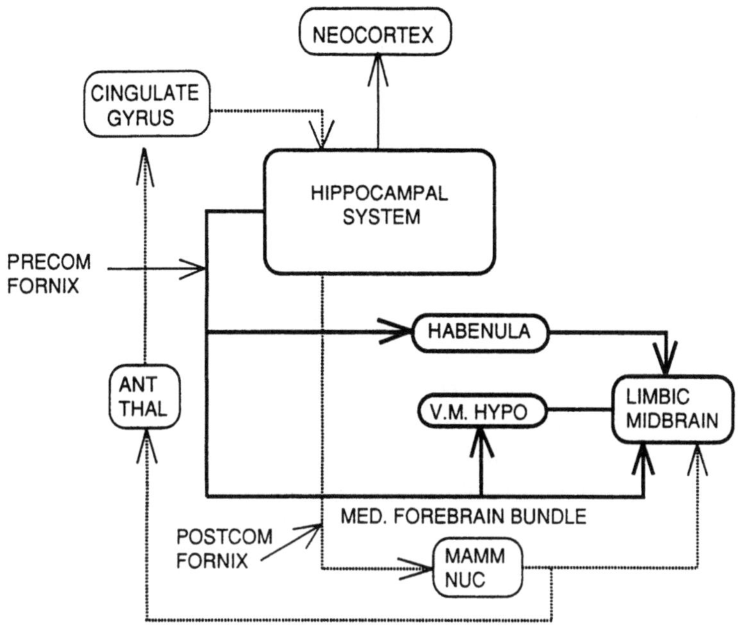


FIGURE 5.2. Simplified view of hippocampal outputs; Papez's circuit is indicated by dotted lines.

There is no solid evidence linking the mammillary nuclei and/or their projection to the anterior thalamus to the hippocampal syndrome. Kim, Chang, and Chu (1967), for example, found massive lesions of the mammillary bodies in dogs to have no detectable effect on delayed alternation. On the other hand, Rosenstock, Field, and Greene (1977) reported mammillary body lesions to significantly reduce learned alternation in the T maze. Their data show, however, that the lesioned rats were alternating at rates far above chance. Combining the delay and nondelay conditions, the lesioned rats averaged only 30 errors (nonalternations) out of 160 trials rather than the 80 expected by chance. The same issue can be raised with regard to a later study by this group (Field, Rosenstock, King, & Greene, 1978), where rats with lesions of the mammillary nuclei or of the two paths leading from them were defective in trials to criterion and yet appear to have been alternating in the high 70 or low 80% range. These studies involved an automatic maze that allowed the animals to make numerous responses within a short time, and interference due to the massing of trials may have been the critical factor. The latter is suggested by the results of the only mammillary study employing a near-traditional SAB procedure. The mice in that study (Beracochea & Jaffard, 1987) were given six to nine trials in succession, and the mammillary-lesioned animals displayed normal SAB on the first pair of trials. The alternation rate was below normal only on those trials after the first pair.

Studies involving mammillothalamic tract lesions have primarily found the animals to be normal or even super-normal on some hippocampus-related tasks.

For example, Krieckhaus, Coons, Greenspoon, Weiss, and Lorenz (1968) found the lesion to have no effect on passive avoidance or shocked T-maze performance. Krieckhaus (1964) found his tract-lesioned rats to be worse than normal in the shuttle box whereas hippocampally damaged rats are superior to normals on this task. Thus, the mammillary nuclei and associated tracts have simply not been shown to be importantly involved in the hippocampal syndrome, and much of the evidence is in the opposite direction. Ordinary two-trial spontaneous alternation was unaffected by mammillary lesions in the only study using an orthodox paradigm. In summary, the evidence as a whole suggests that the Papez circuit is not the place to look for SAB or (in people) dramatic deficits in recent memory.

Turning to specifically human evidence, it appears to be a well-kept secret that many people with lesions of the postcommissural fornix have been found to have no obvious memory deficit. Squire and Moore (1979) located 47 cases in the literature in which the patient had been subjected to surgical interruption of the fornix, and in only three of these cases was a memory deficit noted. They suggest that in the cases where no memory deficit was found the lesion was confined to the postcommissural fornix. In the three cases with a loss of recent memory the whole fornix (pre and post) was involved. It should be noted that in animal studies the entire fornix is usually severed at a point adjacent to the hippocampus. In the one alternation study in which the postcommissural fornix was selectively damaged the rats did not display the hippocampal syndrome (Johnson et al., 1977). The evidence presented by Beracochea and Jaffard does suggest, however, that people with mammillary lesions (or postcommissural fornix damage) might prove to be especially prone to interference effects on memory.

### *Dorsomedial Nucleus and Frontal Lobes*

But what about the dorsomedial nucleus? As mentioned earlier, there is nothing about this nucleus that would lead one to expect it to be involved in memory and/or alternation, and yet there are several reports of memory deficits in people with dorsomedial damage (e.g., Cramon, von Hebel, & Schuri, 1985; Kritchevsky, Graff-Radford, & Damasio, 1987; Speedie & Heilman, 1982, 1983; Winocur, Oxbury, Roberts, Agnetti, & Davis, 1984). Most of these reports involve only one or two cases, and the total is not large. Studies involving monkeys appear to support the idea of a link between this nucleus and memory (e.g., Aggleton and Mishkin, 1983a, 1983b; Isseroff, Rosvold, Galkin, & Goldman-Rakic, 1982; Zola-Morgan & Squire, 1985). The lesions typically include structures such as anterior thalamus, mammillothalamic tract, etc., in addition to the dorsomedial nucleus. The Cramon group had two patients in which the lesion was confined to the dorsomedial nucleus, as did the Kritchevsky group. All four of these patients had perfectly normal memory. Graff-Radford, Tranel, and Brandt (1988) reported another patient with bilateral dorsomedial lesions whose memory was normal after recovery from the immediate effects of the stroke. All three groups speculated that memory deficits were probably due to damage to some structure other than the dorsomedial nucleus, with the mammillothalamic tract being one possibility. It is, however, far from clear that this tract is importantly involved in the mnemonic functions of the hippocampal system.

I now suggest another possibility. The habenula is virtually embedded in the dorsomedial nucleus and thus very likely to be included in any dorsomedial lesion. Moreover, it forms a link in one of the powerful downward projections of the hippocampal system. Amazingly, not one investigator has yet studied the possible effects of habenula (or habenulo-interpeduncular tract) lesions on SAB. Reports of the effects of dorsomedial lesions on alternation appear, on the surface, to be contradictory. Means, Harrell, Mayo, and Alexander (1974) and Weis and Means (1980) found a complete abolition of SAB in rats with dorsomedial lesions, along with other hippocampal syndrome effects such as resistance to extinction and a lack of habituation. Tigner (1974), on the other hand, found dorsomedial lesions to have only trivial effects on SAB and reversal learning. However, the sizes and locations of the lesions appear to have been different. The very large lesions in the studies of Means and his coworkers were discovered to have unintentionally included the habenula. The Means group found, in fact, that the deficit in alternation was highly correlated with the amount of habenula rather than dorsomedial damage. Another hippocampus-dependent behavior, DRL or delayed bar press responding, is also significantly impaired after habenula lesions (Evans & Thornton, 1984). I conclude that while lesions, which include the dorsomedial nucleus, might knock out alternation and/or human memory, the nucleus itself is only incidentally involved. A "bandwagon effect" seems to have resulted in a tendency to refer to extensive anterior thalamic damage as a dorsomedial lesion. A case in point is Squire and Moore's (1979) attributing the amnesia of the famous patient "N.A." to a dorsomedial lesion when the damage was produced by a sword going up N.A.'s nose into his brain.

At this point we will return to the frontal lobes, the main projection region of the dorsomedial nucleus. Thousands of people have been subjected to some variation of prefrontal lobotomy, and these people have not been reported to have anything like the recent-memory loss associated with hippocampal system damage. There are, however, occasional claims of defective performance on learning or memory tests in people with frontal lobe damage. There have also been numerous claims of memory deficits in monkeys with frontal damage going back to Jacobsen's classic study in 1935. Stuss and Benson (1983) conclude from their analysis of the human cases that frontal lobe damage does not produce true amnesia but that frontally damaged people might perform poorly on some memory tests because of tangential deficits such as a tendency to perseverate or an inability or unwillingness to follow rules and/or formulate plans.

There is now evidence suggesting that the human frontal lobes might be involved in judging the temporal order or the frequency of repeated events (Milner, Petrides, & Smith, 1985; Smith & Milner, 1988). In the latter study lesions of the right frontal lobe were associated with a deficit in remembering how many times a given stimulus had been presented, but the frontally damaged subjects could remember that they had seen the stimuli. Olton, Wenk, Church, and Meck (1988) compared rats with frontal or hippocampal-system lesions on a temporal discrimination task and found only the latter animals to have a defective "working memory." In light of this evidence we might speculate that frontal lesions could conceivably reduce SAB when many massed trials are used. The critical feature of the typical delayed alternation task may be the use of highly massed trials, with the monkey being unable to distinguish between what it did on the previous trial from what it did on the trial preceding the previous one, etc. Meanwhile, the evidence as a whole argues against the frontal lobes in general being involved in SAB or conscious memory in any direct way.

As mentioned earlier, Divac et al. (1975) did find frontal (specifically medial) lesions to reduce alternation. Only one other study (Kinghorn, Fleming, & Anderson, 1987) appears to have found frontal lesions to reduce SAB. That study failed, however, to report what the alternation rate was, used dorsal rather than medial lesions, and reported that unilateral and bilateral lesions did not differ in their effects on SAB. The latter is definitely eyebrow-raising, and one suspects that turning biases might have affected the results. In contrast to this pair of reports, many other studies have found frontal lobe lesions to have no affect on alternation (Douglas, 1975; Holson, 1986; Markowska & Lukaszewska, 1974; Nonneman & Kolb, 1979; Weis & Means, 1980). One can conclude with a fair degree of certainty that the vast bulk of the frontal cortex has nothing to do with ordinary SAB, though it might be involved in massed-trial alternation of either the learned or spontaneous variety. It also remains possible that some small fragment *might* be related to ordinary two-trial alternation. That fragment is the midline region, powerfully innervated by the hippocampus, that receives a dopaminergic input from nucleus A10. Taghzouti, Louilot, Herman, LeMoal, & Simon (1985) found lesions of nucleus accumbens to abolish alternation, but they noted that the lesion also knocked out the dopaminergic innervation of the adjacent medial frontal region. Finally, this fragment of cortex also includes the region that the late Robert ("Mac") McCleary termed "subcallosal cortex" and which he believed mediated inhibition. In summary, it remains possible that dopamine-innervated midline frontal cortex might be involved in alternation but that more evidence will be required before this can be established with any degree of certainty.

It is important to recognize the supreme difficulty in extrapolating from animal memory research to human. As mentioned earlier, people with a recent-memory loss retain an ability to learn some tasks. Conversely, people with perfectly good memory find it difficult to learn some problems. In animal research, however, there is a strong tendency to say that amnesia does or does not exist on the basis of performance on a task that in some cases appears to have been arbitrarily chosen. It is to be hoped that the results of a recent study by Freedman and Oscar-Berman (1986) might cure some of us of this tendency. They used two common "monkey memory" tasks, delayed response and delayed alternation, but gave the tests to people with frontal lesions or the amnesic syndrome instead of to lesioned monkeys. People with bilateral frontal damage performed poorly on both tasks despite having normal memory on standard memory tests and in real life, and there proved to be no relationship between performance on the monkey tasks and on human memory tests. Because of these findings we now know that it is invalid to infer amnesia from the fact that an animal might perform poorly on these tasks.

*Electroconvulsive Shock*

Before leaving the subject of human memory I will cite still another case in which a procedure that powerfully affects human memory also knocks out SAB. Electroconvulsive shock (ECS) produces a temporary human version of the hippocampal syndrome (Zamora & Kaelbling, 1965) and by the present line of reasoning might be expected to knock out SAB and produce other facets of the rat hippocampal syndrome. Indeed, three different studies have found ECS to produce a severe and prolonged impairment of SAB (Dokla, Kasprow, Sideleau, & Boitano, 1980; Douglas, Pagano, Lovely, & Peterson, 1973; Keyes, 1973). References in these papers show that ECS also produces a full range of hippocampal syndrome effects in addition to the alternation deficits. Thus, I believe that we are on the right track when we suspect that lesions truly abolishing alternation in rats will affect human recent memory, and vice versa. So far the evidence strongly supports this viewpoint, and if the relationship holds up, we can predict memory deficits from lesions of such unlikely places as the limbic midbrain and the ventromedial nucleus of the hypothalamus, to be discussed next.

*The Limbic Midbrain and Other Deep Places*

As mentioned earlier, the major descending pathways from the hippocampal system tend to converge in the limbic midbrain, a region that is anatomically pure pandemonium. It contains such potentially important regions as the nuclei A10, interpeduncularis and median raphe, plus numerous ascending and descending pathways. Lesions in the ventral midbrain are thus anatomically difficult to analyze, but they appear to produce an unambiguous loss of SAB. Galey, Jaffard, and LeMoal (1976, 1979) found lesions located in the general vicinity of the interpeduncular nucleus to completely abolish alternation. These lesions also produced increased activity and hyper-reactivity (LeMoal, Stinus, & Galey, 1976), severe defects in reversal learning (Galey et al., 1979), an impairment in passive avoidance (Wirtshafter, 1981) and in extinction (LeMoal, Cardo, & Stinus, 1969), as well as superior learning of the shuttle-box problem (LeMoal et al., 1969; Wirtshafter, 1981). This lesion appears to produce a full-blown hippocampal syndrome. The only problem is *why*. The LeMoal-Galey group has identified nucleus A10 as a likely culprit. Dopaminergic neurons in this nucleus ascend to innervate the ventral portions of the striatum (including nucleus accumbens), the medial frontal lobe, lateral septum, and amygdala. Taghzouti et al. (1985) found 6-OHDA lesions of nucleus accumbens to abolish SAB, and this certainly appears to support the preceding idea as well as Isaacson's (1984) long-standing suspicion that the hippocampus exerts its effects upon this pathway. On the other hand, Wirtshafter's lesions were tiny knife cuts intended to sever selectively the interpedunculo-tegmental tract and unlikely to have damaged A10. But this portion of the brain is a veritable jungle (see Herkenham & Nauta,

1979; Morley, 1986), and considerable research will be required before the relationship between structure/pathway and function can be untangled. One distinct possibility is that the lesion affects hippocampus-related behavior by destroying an ascending nondopamine pathway. Montone, Fass, and Hamill (1988) found neurons located atop the interpeduncular nucleus to project forward to the septal region and hippocampus. The cholinergic pedunculopontine nucleus also projects through this area on its ascending path to the equally cholinergic diagonal band nucleus (Semba, Reiner, McGeer, & Fibiger, 1988).

In summary, numerous elements of the hippocampal syndrome, including a complete loss of SAB, are produced by lesions of the ventral midbrain. Unfortunately, we have no idea what the effects of such lesions on human memory might be. Memory defects have been noted in many people with midbrain lesions (e.g., Casteigne et al., 1981), but the patients invariably have additional damage. Although some evidence cited herein suggests that the lesions might act by damaging the ascending dopaminergic pathway from nucleus A10, one then wonders why catecholamine-related drugs are so spectacularly unsuccessful in affecting SAB when employed at reasonable dosage. As we have seen, there are just too many possibilities and too little evidence for any authoritative pronouncements to be made on this topic at this time.

This hippocampal portion of the analysis will be concluded with a finding that at first glance might appear quite strange. Singh (1973) found lesions of the ventromedial nucleus of the hypothalamus to have effects similar to those of septal lesions on SAB, reversal learning, and passive avoidance. Both lesions completely wiped out alternation, with the rates being slightly below 50%. It is unfortunate that the anatomical connections of the ventromedial nucleus (aside from its powerful input from the amygdala) are so obscure, though after a synapse or two the precommissural fornix system can reach it (see Nauta, 1960). What I find to be most interesting is that the ventromedial nucleus has been postulated to be inhibitory on the basis of both electrophysiological observations (Kaada, 1951) and behavioral evidence (Stellar, 1954). If so, and if the hippocampal system is related to this nucleus, then it follows that the hippocampus must excite or "turn on" the ventromedial nucleus. Although attempts to link the hippocampus to inhibition go back at least three decades (e.g., Grastyan, 1959), no one has suggested that hippocampal output axons are directly inhibitory. To the contrary, the axons of hippocampal pyramidal neurons probably secrete acidic amino acids, and thus are likely to excite the next cell (Cotman, Monaghan, Ottersen, & Storm-Mathisen, 1987). Hippocampal-mediated inhibition might thus consist of turning on inhibitory subregions and/or inhibitory interneurons within subregions.

In any event, there are two highly relevant single-case reports in the human literature (Alpers, 1937; Reeves & Plum, 1969). Both patients had tumors in or near the ventromedial nucleus, and in both cases normal ventricular pressure and/or appearance reduced the probability of pressure effects on remote structures. Alper's patient became exceedingly irritable and uninhibited, displayed spatial disorientation (e.g., drove on the wrong side of the road), and had a severe recent-memory deficit. The patient of Reeves and Plum was also extremely irritable, aggressive, uninhibited, and grossly obese. Periods of normality were interrupted by periods of disorientation and severe recent-memory loss, with the latter presumably becoming a permanent condition in the terminal dementia. It would be difficult to imagine a closer parallel to the behavior of rats and cats with ventromedial lesions. One is tempted to "explain away" the severe memory deficit by saying that the tumor probably encroached upon the post-commissural fornix. The problem, however, is that we have just seen that damage to the human postcommisural fornix probably does not produce a deficit of recent memory in people or the hippocampal syndrome in animals.

At this point you may be asking, "What does it all mean?" First, it means that SAB is securely associated with the hippocampal system. Second, it means that when we study alternation we are studying the epitome of what the hippocampal system does. Third, it means that when we study the hippocampal system we are studying a mechanism crucially involved in SAB. What we cannot do at this point is decide which of several critical factors in SAB can be attributed to the hippocampus. We would be foolish if we even tried, considering that there is another giant chunk of the brain that is also crucially involved in alternation. It includes the vestibular system and brain regions, such as the corpus striatum, to which this system powerfully projects.

## *The Second Story: SAB and Spatial Orientation*

### *Alternation and Spatial Neglect*

In Table 5.1 you will see two lonely rats with caudate nucleus lesions averaging zero alternation. They were a part of a study in which I was testing alternation in rats that had been operated upon by various students who were in the process of learning surgery. The two rats in question were supposed to have bilateral hippocampal lesions, and I expected them to average about 50% alternation. Instead, again and again they entered the same alley. One rat always turned to the right while the other persisted in turning left. Each animal was eventually given 15 tests, and neither alternated even once. If I had known that these animals had damage to the caudate nucleus, I would probably have viewed this behavior in terms of a motor deficit. As an unbiased observer, however, I could not help feeling that each animal failed to acknowledge the existence of one of the alleys. The first rat, for example, never even glanced into the left alley before making any of its 30 responses, while the second completely ignored the right alley. Both were enthusiastic maze runners exhibiting no symptoms of fear.

Histology later revealed the right-turner to have massive right caudate damage while the left-turner had a large left caudate lesion. In both cases there was some unilateral hippocampal damage, but this by itself could not possibly explain the behavior. At that point in history (early 1960s) the zeitgeist was not conducive to viewing rats as having attentional deficits, but a liberalization of thinking occurred during the 1970s, when contralateral sensory neglect was accepted as an explanation for the behavior of rats with unilateral lesions of the lateral hypothalamus (Marshall & Teitelbaum, 1974). Sensory neglect was later found to be due to the lesions having severed the dopamine pathway from substantia nigra to the corpus striatum (Marshall, Richardson, & Teitelbaum, 1974). The corpus striatum (caudate, putamen, accumbens, olfactory tubercule) appears to be incapable of normal functioning when deprived of its dopaminergic input, though it can compensate for a partial loss.

It is clear that if rats with unilateral hypothalamic lesions and/or severed nigrostriatal tracts were tested for SAB (they have not been), they would certainly score close to zero, because they have been frequently reported to turn virtually always in the same direction. But is it really fair to infer from perseveration that the rat lacks alternation? In a word, no – not without an intensive analysis of the data (see Douglas, 1966a). In a simple and straightforward mathematical model of SAB the observed alternation rate becomes reduced when a preference for one of the two alternatives increases even though the underlying "urge" to alternate is assumed to remain constant. The details of the calculations can be found in the preceding study, and they show that if rats alternate 85% of the time when there is no bias for either alley and/or turn, then the observed rats drops to below 45% if one alley is nine times as attractive as the other, assuming a constant urge to alternate. If one alley or one turning response is infinitely preferred to the other, then you would see a rate of 0% alternation in rats that possessed a normally strong motive to alternate. Thus, we simply cannot claim that unilateral caudate or nigrostriatal tract lesions knock out SAB if "alternation" is taken to mean the underlying process. The interaction between bias and alternation is illustrated in the study of Salamone, Beart, Alpert, and Iverson (1984). They found rats with unilateral lesions of nucleus basalis – in a region *not* projecting to hippocampus – to be impaired in reinforced T-maze alternation and to possess strong side biases. The correlation between bias and alternation was −0.97. This is, however, an extremely complex theoretical issue and it is entirely possible that the mechanisms that are knocked out in the animal with sensory neglect are normally involved in SAB. Alternation is, in fact, a dynamic neglect process.

As mentioned earlier, an alternating rat must have a sense of spatial orientation, and sensory neglect is one of many kinds of spatial defect. As we shall see, people with an inability to navigate in space do not necessarily display anything so dramatic as sensory neglect. Sensory neglect associated with cortical lesions is usually temporary, fading away into a condition in which one side of space is favored over the other (the "extinction" phenomenon discussed earlier). Neglect is associated with right hemisphere damage often described as "parietal" but in fact including the superior temporal lobe (see Vallar & Perani, 1986). Nearly all features of neglect can be explained by the hypothesis that it is due to inactivation or depression of the corpus striatum. Some of the evidence follows.

First, it has been known for many decades that animals with unilateral damage to the corpus striatum "turn towards the lesion" or neglect contralateral space. Neglect can also be produced by damage to any of the major inputs to the corpus striatum. This includes the dopamine input mentioned earlier and damage to the thalamic reticular formation or to the cortical regions that most heavily innervate the striatum (parietal and frontal lobes). Sensory neglect due to unilateral frontal lesions can occur in monkeys or people but has not to my knowledge been seen in nonprimates. In primates a frontal region including the eye fields appears to be involved in directing orienting behavior to contralateral space, but rodents do not appear to "point" their eyes and may have only a rudimentary development of frontal orienting systems. Research using the 2-deoxy glucose technique has established that cortical lesions that cause neglect strongly depress subcortical regions such as the striatum and thalamus (Duel & Collins, 1983). Finally, neglect has been seen in people with lesions that include the corpus striatum, but spare cortex (Healton, Navarro, Bressman, & Brust, 1982) and in people with damaged striatum-related thalamic regions (Motamura, Yamadori, Ogura, Sakai, & Sawada, 1986).

Some investigators have considered neglect to be merely an inability or unwillingness to look at the neglected half of space, or as a mechanical rather than a cognitive defect. It is true that people with left-space neglect are reluctant for their eyes to cross the midline to the left visual field, but they also neglect the left side of things presented in the "good" right field. In people, unlike animals, neglect of the right half of space is extremely rare, and this is explainable in terms of Mesulam's (1981) model in which the right hemisphere directs attention to both halves of space whereas the left hemisphere directs attention only to the right. This suggests that the human right corpus striatum can direct attention to either half of space whereas the left striatum is limited in its spatial functioning. Bowen, Hoehn, and Yahr (1972) found spatial deficits in humans with either bilateral or right hemisphere damage to the corpus striatum, as inferred from Parkinsonism symptoms.

Several studies have temporarily "cured" neglect by forcing the eyes into the neglected field via vestibular stimulation (Cappa, Sterzi, Vallar, & Bisiach, 1987; Rubens, 1985; Silberpfennig, 1941). Cold water in the left ear forces the eyes to move left, as does warm water in the right ear. In the studies just listed this caloric vestibular stimulation resulted in a dramatic restoration of the lost field. This evidence might appear to support the "response" theory, but several ingenious studies have shown that the defect is psychological rather than merely mechanical and that it involves attention as well as orienting.

Bisiach and Luzzatti (1978) had two of their left-neglect patients try to imagine Cathedral Square in Milan as seen from the Cathedral and as seen looking towards the Cathedral. In both cases they strongly tended to remember things to the right and omit things to the left, so that a given feature was or was not remembered depending on the mental viewpoint. In a similar study Meador, Hammond et al. (1987) had a left-neglect patient imagine his home as viewed from the front or back door, and he remembered things to the right when "looking" from the front door and omitted items to the left. From the back door the set of things he remembered or imagined was reversed. People habitually think of the west as being left, and when left-neglect people are asked to name the states they tend to include most eastern states and forget the western (Morrow, 1987). Thus, in cases of sensory neglect some unknown mechanism actively interferes with imagery and/or memory retrieval in one hemifield. This evidence also demonstrates that the vestibular system can control the direction of attention. We can explain these facts by postulating that the vestibular system activates the corpus striatum, thus compensating for other lost inputs.

*Navigation*

There is not sufficient space to fully discuss the variety of spatial defects associated with posterior neocortical lesions (see Critchley, 1966). There is, for example, a condition in which people get lost because they cannot recognize landmarks, but this may be analogous to prosopagnosia (failure to recognize faces) and not truly a navigational defect. It is unfortunate that the deficit most relevant to our concerns has been the least studied in formal experiments, though perhaps the most noted informally. This is the lack of something that could be called "navigational ability." The latter involves the use of movement-generated feedback and the creation of an internal spatial cognitive map. The possession of such a map can be demonstrated by the taking of shortcuts.

Animal and human research shows that there are many ways of guiding behavior. A "scene" is hereby defined as the sensory experience occurring with the individual in one location. Within a scene a person can get around by using egocentric localization, or by responding to stimuli in the here and now. A series of scenes can be linked up via verbal crutches or by a chain of conditioned associations (see Beritoff, 1965). But these methods demand that the person or animal laboriously follow one specific pathway from A to Z. Navigation, however, allows the animal to go straight from A to Z without passing B,C,D, etc. I will argue that navigation is involved in SAB and that navigation involves the vestibular system. My plan for the following portions of this chapter is to begin with a set of structures that are strongly related to the vestibular system, show that they are involved in SAB, and then demonstrate that they are also involved in navigation.

*Alternation and the Greater Vestibular System*

The "greater vestibular system" is hereby defined as the immediate vestibular system (labyrinth, brainstem vestibular nuclei) plus all brain regions powerfully innervated by this system (see Figure 5.3).

Unlike other sensory systems, the vestibular system is impure right from the very beginning. The first relay nucleus in the visual system (the ganglion cell layer of the retina), for example, is 100% visual. In contrast, the first-order vestibular relay nuclei receive proprioceptive and visual inputs that interact with the vestibular. For example, the very same neuron that fires when a rat is moved forward in the dark will also fire if the animal is held still and the visual environment moved towards the animal (Horn & Miller, 1982). Further, if the vestibular system is to monitor movements of the whole animal in space, then it must employ proprioceptive inputs from the neck and other muscles to cancel out

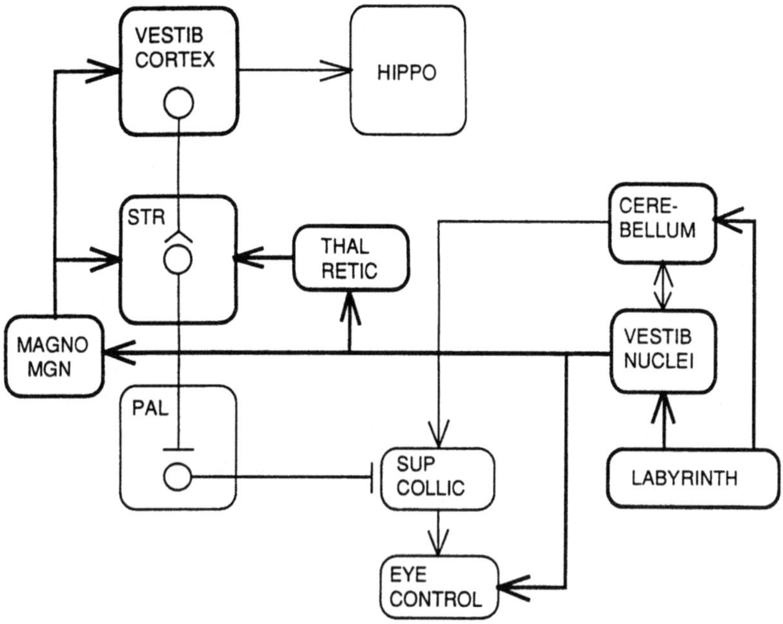


FIGURE 5.3. Some ascending vestibular connections as described in the text. Arrows refer to "flow of information," forks indicate excitatory synapses, bars inhibitory. STR = corpus striatum; PAL = pallidum; MAGNO MGN = magnocellular portion of medial geniculate nucleus.

vestibular stimulation produced by irrelevant head movements. Perhaps this vestibular-visual-proprioceptive conglomeration ought to be named the "spatial system."

It is impossible to separate the vestibular system from the cerebellum. First-order vestibular neurons innervate the cerebellar vermis (Kotchabhakdi & Walberg, 1978), and the vestibular nuclei have profuse interconnections with the cerebellum (see Pompeiano, 1974). Electro-physiological studies reveal a powerful projection from the vestibular nuclei, probably by more than one pathway, to widespread regions of the corpus striatum (Abraham, Copack, & Gilman, 1977; Copack, Dafny, & Gilman, 1972; Potegal, Copack, deJong, Krauthamer, & Gilman, 1971). One pathway is the anatomically verified projection to the thalamic reticular formation (Cornwall & Phillipson, 1988; McGuiness, Dalsass, Proshansky, & Krauthamer, 1976), which is a major supplier of striatal input. A second pathway appears to take an ascending route medial to the lateral (auditory) lemniscus to the large-celled part of the medial geniculate nucleus. The latter appears to project directly to the striatum (Veening, Cornelissen, & Lieven, 1980) as well as to the anterior suprasylvian cortex (in carnivores). The latter in turn projects down to the corpus striatum. There is some question about the exact location of the cortical vestibular projection(s) in primates. Frederickson, Figge, Scheid, and Kornhuber (1966) localized the primary vestibular projection in the lower parietal lobe in monkeys, whereas a more recent study using human subjects tentatively places it just behind auditory cortex in the temporal lobe (Friberg, Olsen, Roland, Paulson, & Lassen, 1985). The latter corresponds to the region where electrical stimulation of the human brain causes sensations of twirling or movement (Penfield, 1957). It should be noted that both areas lie within the cortical region typically damaged in people with sensory neglect (Vallar & Perani, 1986).

The next question we will ask is whether SAB is dependent upon the integrity of the greater vestibular system. First, purely behavioral evidence, as reported in Chapter 2, links alternation to the vestibular system. Rats alternate at very high rates if each of the two trials is given in a physically different maze, even when each maze is in a different room (Douglas, 1966a). This is true, however, only when the two mazes are oriented in parallel (e.g., both pointed north). When the mazes are oriented at right angles (e.g., one pointed north, the other east), alternation does not occur even if the two mazes contain similar multisensory internal decorations. Second, if rats are tested in Y-type mazes with various angles between the choice alleys, then the alternation rate depends on the angle through which the head turns, and it falls to about 50% with an angle of 45 degrees or less (Douglas, Mitchell, & delValle, 1974). Third, if rats are disoriented by vigorous back-and-forth turns in a dark room between trials, then they do not alternate in two parallel mazes. It should be noted that while there is some evidence that superficially appears to contradict all of the findings listed earlier (see Potegal, 1982), the "contradictory" studies (1) fail to replicate the conditions of the experiment, (2) have other shortcomings such as a 33% balk rate, or (3) have themselves been contradicted. As we shall see later, conclusions based on the work cited herein have been resoundingly confirmed in other species in other circumstances.

Now let us turn to the lesion data. First, rats with vestibular malfunction due to middle ear disease do not alternate (Douglas, 1966b; Franken & Baker, 1969). Bilateral destruction of the vestibular nerve abolishes SAB (Potegal, Day, & Abraham, 1977), as do bilateral lesions of the vestibular nuclei (Potegal, 1982). Finally, Douglas et al. (1979) found genetic defects of the vestibular static organs to abolish SAB in some mice. This is somewhat surprising, since one would imagine the semicircular canals to be more critical than the static organs, but Potegal (1987) found both static organ and semicircular canal lesions to significantly impair spatial orientation. There would appear to be little doubt that alternation is abolished by lesions of the labyrinth or the vestibular relay nuclei.

The next question is whether the cerebellum is involved in alternation. Probably the most definitive study is that of Pellegrino and Altman (1979), because their lesions were restricted to the cerebellum. The vast majority of small neurons in the cerebellum are products of postnatal cell division in rats and mice, and virtually all of these can be selectively killed by repeated X-ray treatments between 4 and 15 days. As was shown in the previous study, this treatment abolishes SAB. Milder X-ray schedules were less effective in killing postnatally dividing neurons and had little or no effect on SAB. Next, there are numerous genetic defects of the cerebellum in mice, and many of these colorfully named defects have now been shown to be associated with a lack of SAB. This includes "reeler" (Bliss & Errington, 1977), "weaver" (Lalonde, 1986), "nervous mutant" (Lalonde, Botez, & Boivin, 1986), "lurcher" (Lalonde, Lamarre, Smith, & Botez, 1986), "Purkinje cell degeneration" (Lalonde, Manseau, & Botez, 1987), and "staggerer" (Lalonde, Manseau, & Botez, 1988). Although each defect usually involves other regions in addition to the cerebellum, there is a great deal of variety both in the nature of the cerebellar defect and in the extra-cerebellar effects. Pending further evidence, one would have to conclude that extensive lesions of the cerebellum abolish SAB. The findings of Pellegrino and Altman (1979) stress the stipulation that the lesion be extensive, as is the case with the genetic defects described earlier.

Turning to the corpus striatum, we find that in 1969 Kirkby reported lesions of the caudate nucleus to abolish alternation. Potegal and Squire (1974) replicated these results, as did Divac, Wikmark, and Gade (1975). Karpiak, Rapport, and Bowen (1974) injected antiserum to synaptic membranes into the lateral ventricles in rats. This produced dramatic changes in caudate nucleus electrical activity (spikes, bursts, etc.) and a loss of SAB. Finally, Taghzouti et al. (1985) found the destruction of dopamine axons in nucleus accumbens to abolish alternation. Two studies have been cited as being contradictory to these, but I find only one of these to be truly inexplicable. Borst, Delacour, and Libouban (1970) appear to have genuinely found caudate lesions to have no effects on SAB. On the other hand, although Baettig (1963) is often cited as showing that caudate lesions do not affect SAB, he used an unusual procedure involving forced trials and goals differing greatly in their attractiveness. When Baettig's results for both goals are combined, his caudate-lesioned rats averaged 52% alternation, as compared to 75% preoperatively, and the drop is statistically significant ($t = 6.3$; $p < 0.01$). Thus, Baettig was ironically the first to discover that caudate nucleus lesions abolish SAB. What is remarkable about these results is that in all cases the lesion was very small in comparison to the relatively huge size of the striatum. It is very doubtful that a lesion of a similar proportional size in either the hippocampus or the cerebellum would have detectable effects once the shock of surgery had subsided.

This discourse should ideally conclude with a demonstration that SAB is abolished by lesions of vestibular cortex. As was pointed out earlier, however, there has been no truly systematic study of possible relations between cortical lesions and SAB, and the rat's vestibular cortex has not yet been identified, much less studied behaviorally. The "posterior association cortex" destroyed by Thomas and Weir (1975) may well prove to be crucial for alternation. They found the lesion to severely disrupt postoperative retention of a Lashley III maze problem, which suggests that this region might be involved in spatial orientation and/or spatial memory. Since vestibular projection cortex has been identified in the cat (Potegal et al., 1971), and since cats do spontaneously alternate (Frederickson & Frederickson, 1979), one wonders why the obvious experiment has not yet been done. In summary, there is considerable evidence that SAB is critically dependent upon the integrity of the greater vestibular system. What remains is to show the reason why this is true.

*Navigation and the Greater Vestibular System*

In recent years several studies have investigated the role of the vestibular system in navigation by employing approach behaviors. These include the hoarding response in hamsters (Etienne, 1980, 1987; Etienne, Teroni, Maurer, Portenier, & Saucy, 1985) and the pup-retrieval motive in gerbils (Mittelstaedt and Mittelstead, 1980). These studies involve one or more circular arenas with numerous potential nest locations around the periphery (48 in the latter case). A pup, for example, can be placed on a small dish in the center of the arena, and the mother runs out to retrieve it, or the animal can be passively transported to the center and shown food to hoard, etc. When the arena is rotated prior to the return trip, the rodents demonstrate spatial orientation by returning to the place in absolute space where the nest was located. The Etienne group discovered highly accurate homing to the spatial location of the nest when all conceivable nonvestibular cues (including proprioception and even magnetic field) were systematically eliminated. Both groups found strong evidence that rodents integrate a series of angular turns on an outbound journey so that they can head straight home by reversing the vector 180 degrees. If the arena is rotated, the vector is adjusted.

Interestingly, hamsters and gerbils (and presumably other rodents) are insensitive to linear sideways displacements, after which their return route is parallel to what it should be. This can be seen in the experiment where Etienne and coworkers "fooled" the hamsters by removing them from the home arena and placing them in an adjacent one such that they could either go to the general absolute location of the nest or head in the direction where the nest would be if they were in the home arena, but not both. In this case they headed off in the correct compass direction, but away from home. I have made extensive (but unpublished) observations of the same phenomenon in alternation. That is, I used a two-maze test, but with the two parallel mazes side by side. If the mouse turned to the right on its first trial in the left-hand maze, then alternation in the right-hand maze would mean returning to the same general spot as on the first trial. In fact, that is exactly what they did, demonstrating a preference for direction over place. In the real world, of course, animals are not picked up by giant primates and jogged over to the side. These studies confirm Barlow's (1964) brilliant analysis of inertial navigation and its parallel in the greater vestibular system. They also show that my studies were biased. That is, while animals can and do employ vestibular-based navigation, the studies show that they also make good use of distal visual cues when these are available and can be used throughout the behavioral episode. In most of my studies I have employed a hooded lamp directly over the choice point in an otherwise dark room, and this may very well have prevented the animals from using distal visual cues. On the other hand, need I remind the reader that most rodents are strongly nocturnal and spend much of their time in underground labyrinths?

Probably the best series of studies on the importance of the vestibular system to navigation is that of Beritoff (1965). He showed that when blindfolded normal dogs and cats were led away from a food dish over a complex path and then released, they ran straight to the dish by a shortcut. If the dish was removed after they had sniffed it, then they ran to the place where the dish had been. Beritoff experimentally and/or surgically eliminated the navigational relevance of all cues except vestibular. Proprioception was eliminated, for example, by his demonstration that the dish could be located after passive transport in a cage. This ability to navigate was abolished by lesions of the vestibular system, by very large (but not small) cerebellar lesions, or by bilateral damage to vestibular cortex. Beritoff also found navigation to be highly defective in people with vestibular malfunction. Unfortunately, Beritoff did not include the striatum in his navigational thoughts, so for this purpose we must turn to some studies by the Potegal group.

In 1969 Potegal made a pioneering use of the eight-arm radial maze for the study of navigation. He tried to train rats with caudate lesions to turn at a constant angle on each trial when started from the ends of different alleys (and thus go to a physically different place on each trial). He found them to be incapable of learning. In contrast, they had no difficulty in learning to go in a given direction as indicated by distal cues. Potegal unfortunately used the term "egocentric orientation" to describe what the lesioned rats lacked. I consider behavior directed by a constant exteroreceptive cue to be "egocentric." But more about that later. Meanwhile, Potegal's group turned to the study of passive transport. Rats were trained to return to a water spout after having been passively transported along a route involving a turn. Return navigation was unaffected by blinding, but lesions of the vestibular nuclei produced a severe and apparently permanent deficit (Miller, Potegal, & Abraham, 1983). Rats with cerebellar cortex lesions were also included in this study, and although they were reported to have made numerous characteristic errors, they were clearly superior to the animals with vestibular nucleus damage. Although details of the cerebellar lesions are not reported, they were obviously not complete, and one can surmise that they were probably not extensive.

A second study by this group examined the effects of caudate and hippocampal lesions on navigation after passive transport (Abraham, Potegal, & Miller, 1983). Caudate lesions severely impaired return navigation while hippocampal lesions did not. Since these were small electrolytic lesions, however, the evidence does not dismiss the possibility of hippocampal involvement. In contrast, both groups were easily able to learn to reach a goal by following exteroceptive (olfactory) cues, as were rats with vestibular lesions in the earlier study.

As we have seen, there is now considerable evidence linking the greater vestibular system with navigation, and we even know some of the details as to how vestibular information is employed. We would, of course, like to see some more human evidence involving the corpus striatum and/or the posterior temporoparietal junction, but this is sparse. Semmes, Weinstein, Ghent, and Teuber (1955) found people with posterior cortical lesions to be highly defective in navigation when required to follow visual or tactile maps. Ratcliff and Newcombe

(1973) compared the effects of posterior cortical lesions on navigation involving whole-body movement (map-following) versus stylus-maze learning. They found performance on the two tasks to be unrelated. Stylus-maze learning was affected by unilateral right-hemisphere damage whereas bilateral lesions were necessary to knock out map following. As they pointed out, a stylus maze is entirely in one place or within what I would call a "scene," map following involves body movement from one place to another. Bowen et al. (1972) found the same severe map-following deficit to occur in Parkinson patients with presumed damage to the corpus striatum.

To my knowledge all other formal studies of spatial orientation deficits in brain-damaged human subjects involve orienting within single "scene" or perceptual frame, and they are thus without relevance here. We will therefore tap into the vein of informal observations. The authors of the map-following studies reported that some of their most defective performers had volunteered the information that they had "no sense of direction," "got lost easily," etc. Other studies also contain similar observations. The following, for example, is a self-description by a Parkinson patient, as cited (p. 488) by Hovestadt, deJong, and Meerwaldt (1987): "I used to walk alone in the wood, fog or no fog, but when the symptoms of Parkinson's disease appeared, I noticed I could not orient myself any more, and in case of fog, I got lost. Now I am too disabled to get lost any more."

The following is a partial sampling (from Critchley, 1966, pp. 337–339) of what relatives, friends, and therapists have observed about various people with posterior cortical lesions: ". . . suffered from an inaccurate conception as to the spatial relationship of objects which were out of her range of vision. Thus, she would imagine that the street lay just outside her parlor (where actually the bedroom was situated) . . . ," ". . . has found herself in the scullery looking for the dining room. Once she arrived at the wrong end of the village looking for the butcher's," ". . . was observed to have great difficulty in finding his way through the hospital . . . and when faced with two or three doors or passages is unable to choose the correct one," ". . . has difficulty in finding her way about, and was recently almost lost on a familiar allotment of which she could not find the gates. In hospital, was noticed to be in constant doubt as to which was her bed," ". . . has no mental image of his own home where he has lived many years, . . . One of his chief problems was disorientation in space and time: e.g., he was unable to find his way around Bellingham—where he had been living for 15 years."

In summary, SAB can be abolished by lesions of either of two separable sets of brain structures. We have good evidence that one set (the greater vestibular system) is involved in spatial orientation, though we are at present unsure of the exact contribution made by each structure in the system. An even greater problem is deciding what the other system (the hippocampal) contributes. My own efforts to solve these problems follow.

## *The Big Picture*

First, the ordinary animal in the course of its daily (or nightly) activities wanders around looking for food, drink, a mate, etc. In the course of this wandering there are many separate locales or "scenes," and seldom can the animal experience, at one time, the entire extent of its journey. The animal could learn the sequence as a linear string, or a "chain-conditioned reflex," but would then be forced to return home by exactly the same route. Instead, the vestibular system (I would argue) records each turn and the brain (striatum) computes the vector for the homeward-bound leg of the journey. The striatum is involved in orienting behavior that is internally directed but not in primary or externally directed orienting. Indeed, cats with near-complete caudate lesions display excessive externally directed orienting such as compulsive following response (Villablanca, Olmstead, Levine, & Marcus, 1978).

The very word "orienting" presupposes a spatial location. In order to generate an orienting response to a stimulus located in a different scene, the striatum must have knowledge of previous movement-generated feedback, or vestibular input. Much of the lesion evidence cited earlier can be explained by the hypothesis that striatal damage prevents the animal from making use of movement-generated feedback. People with navigational defects are commonly reported to be especially helpless in the dark or in a fog. Normal pigeons can learn to peck at a target in the dark, but this ability is abolished by striatal lesions (Mitchell & Hall, 1987). The striatum, of course, acts through its connections with the pallidum, and pallidal lesions prevent rats from using body-feedback control of posture (Labuszewski, Lockwood, McManus, Edelstein, & Lidsky, 1981). But this is getting ahead of the story. We can explain how the brain produces alternation behavior only in terms of a relatively explicit model, and such a model (currently in preparation) would require a book-length monograph for its exposition. Thus, I will be forced here to present a very sketchy version that omits a number of important details.

### *The Striatal System*

We can begin by considering the corpus striatum as an anatomical-physiological entity. Documentation for much of what follows can be found in Nauta and Domesick (1984), Graybiel (1984), or other chapters in the same publication. In brief, the striatum receives major inputs from neocortex (especially parietal and frontal lobes), the thalamic reticular formation, pars compacta of substantia nigra, hippocampus, amygdala, and the dorsal raphe nucleus. Downward projections from neocortex are topologically organized, though with a high degree of overlap, and they are all excitatory and primarily glutamic acid-mediated. Since the parietal lobe is such a major contributor to striatal input, and since so many parietal neurons fire at high rates to attended stimuli, we should not be surprised to find striatal neurons to be attention-related. Buser, Ponderoux, & Mereaux (1974) found many striatal neurons that did not change their firing rates during movements, but instead fired in response to external stimuli. They noted that these stimuli had one common feature, which was that they attracted the animal's attention. On the basis of logic rather than empirical proof, I postulate that all portions of the striatum are interconnected via a vast neuronal chain that mediates lateral inhibition. Rapidly firing neurons in an attended engram then capture

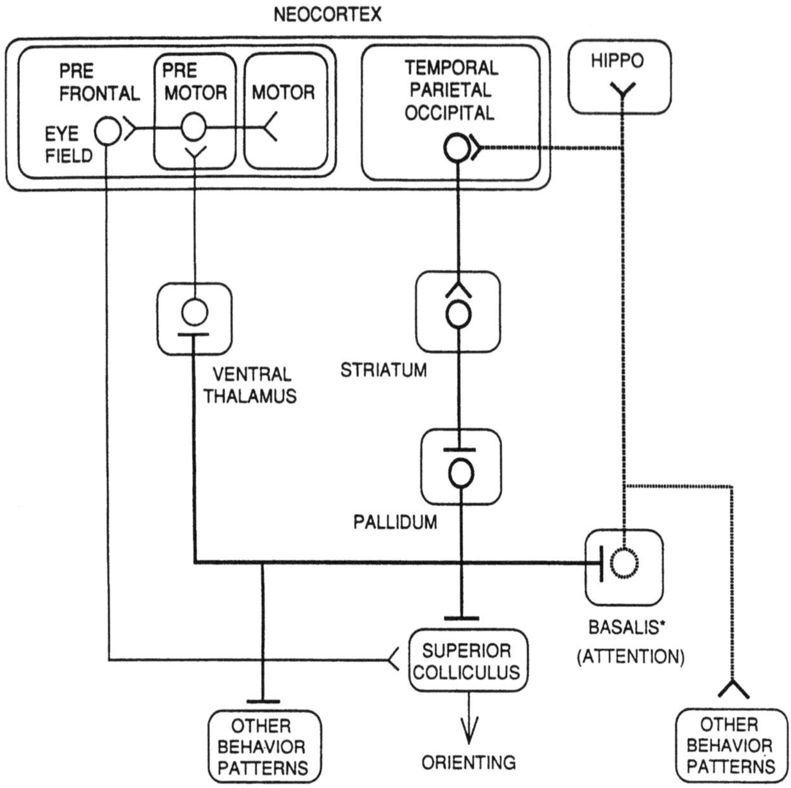


FIGURE 5.4. Some connections of the striatal ("extrapyramidal") system. The term "basalis" refers to long-axon ACh neurons, and synapses are not shown along the path. "Ventral thalamus" means nuclei VLa and VMp while "premotor cortex" means area $6a\alpha$. Forks indicate excitatory synapses, bars inhibitory.

or dominate the striatum and by way of lateral inhibition suppress reactions to less active inputs. This allows us to pay attention to one thing at a time.

The major output neurons of the striatum are GABA-secreting medium spiny neurons that innervate (and inhibit) the pallidum, hereby defined as the external globus pallidus, the entopeduncular nucleus (internal globus pallidus), the so-called ventral pallidum associated with accumbens and the olfactory tubercule, and pars reticulate of the substantia nigra. The latter is not to be confused with pars compacta, which is not part of the pallidum. The output neurons of the pallidum are also GABA-ergic and inhibitory, and they participate in two different but possibly related functions illustrated in Figure 5.4.

First, the pallidum inhibits neurons in the ventral thalamus that project to the premotor region of the frontal lobes. Premotor frontal regions are associated with motor planning and are adjacent to prefrontal regions associated with orienting behavior (e.g., frontal eye fields). This pathway has been postulated to mediate the enabling of motor movements rather than their direct production

(Neafsey, Hull, & Buchwald, 1978). By way of this pathway any portion of cortex has access to motor movements. In addition to this motor loop the striato-pallidal system also has a direct output mediated largely, but not exclusively, by pars reticulata of the substantia nigra. Pars reticulata (and the pallidum in general) contains rapidly firing tireless neurons that suppress what I call "behavioral control centers" such as the primordial orienting center located in deep superior colliculus. The latter is our chief concern, and details of how it works are illustrated in the paper by Wurtz and Hikosaka (1986).

The deep superior colliculus is a multisensory region controlling contralateral orienting behavior of eyes, ears, and body, and the fast-firing pallidal GABA neurons tonically suppress this region. A leftward orienting response occurs when the pallidal neurons of the right hemisphere suddenly stop firing, because of their being inhibited by neurons of the right striatum. The striatal neurons become activated when the right cortex displays an interest in something to the left of the animal. One thing that is not entirely clear now is how the striatum and/or parietal lobe employ movement-generated input to revise and update spatial vectors, though there is little doubt that they do. An important feature of the model is a connection that is not yet proven, but which I predict will be established in future research. That is, when the pallidum releases orienting behavior it also releases cholinergic neurons mediating attention. This could be accomplished by connections with pedunculopontinus and, via the latter, to basalis. Or the connections could be directly from the pallidum to basalis, as in Figure 5.5. Or it could be both. In this way a single pallidal neuron simultaneously controls orienting and attention, and yet it is still possible to separate the two functions. Several recent studies have implicated the cholinergic projection to cortex in attention-related phenomena. It is well-established that the "P3" wave (or P300) in a sensory-evoked potential is related to attention. The P3 wave is greatly reduced in people with Alzheimer-type dementia and presumed loss of cholinergic neurons in nucleus basalis (Meador, Hammond, et al., 1987). It is also reduced or abolished by anticholinergic drugs (Hammond et al., 1987; Meador, Loring, Adams, Patel, & Davis, 1987).

### *The Role of the Hippocampus*

The next thing to consider is how the hippocampus fits into the picture, but the problem is an embarrassment of riches. There are in fact numerous hippocampal/striatal interfaces. First, although it has long been thought that the hippocampus projects to just one tiny corner of the striatum (nucleus accumbens), it now appears that the hippocampal system (including subiculum and entorhinal cortex) actually projects to a long strip that includes large portions of the caudate-putamen and largely coincides with the amygdala projection (Groenwegen, Vermeulen-Van der Zee, Kortschot, & Witter, 1987; Sorensen & Witter, 1983). It follows that if the hippocampus inhibits the striatal circuitry it must thereby inhibit orienting/attention, and this by itself would explain SAB. Hippocampal stimulation has mixed immediate effects on nucleus accumbens neurons (Yang &

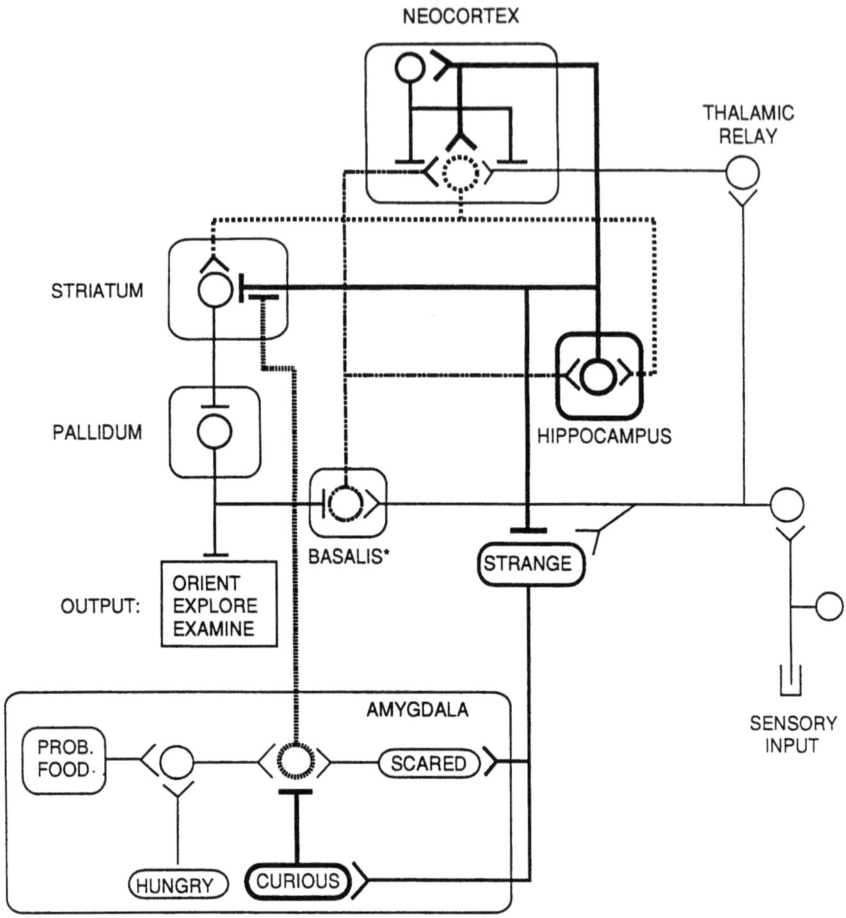


FIGURE 5.5. Highly schematic model of brain mechanisms suggested to underlie spontaneous alternation. Redundant synapses are omitted so that, for example, hippocampal efferents shown as inhibitory could equally represent excitation of an inhibitory interneuron. Again, "basalis" is used very loosely, synapses between cholinergic neurons omitted.

Mogenson, 1984), but what we are concerned with is the effect of such an input on the system. That is, the nucleus accumbens appears to excite or "turn on" locomotion, and *if* the hippocampus excited these circuits, then hippocampal stimulation would produce forward locomotion. In fact, there is no motor response consistently produced by hippocampal stimulation in man or beast. I have personally stimulated electrode sites throughout the hippocampus hundreds of times in unrestrained cats and have seen no evidence of any response at all, with a minor exception that the animal sometimes stops moving when stimulated. Vanegas and Flynn (1968) also found hippocampal stimulation to produce no overt motor behavior but to strongly inhibit motor movements normally elicited by cortical stimulation.

As a second possibility, the thalamic reticular formation (midline and intralaminar nuclei) provides a powerful striatal input. Adey, Segundo, and Livingston (1957) demonstrated that stimulation of the hippocampal system can completely inhibit transmission between brain stem and thalamic reticular nuclei. Third, we have shown that the hippocampal output via the habenula may be critical in mediating SAB. The habenula is also a major target of the pallidum, or a potential site of striatal/hippocampal/amygdala interaction. Fourth, the dopaminergic neurons of pars compacta of the substantia nigra "turn on" the striatum, and future research may reveal that the hippocampal system "turns off" these neurons. Fifth, the hippocampus projects profusely out "backwards" to widespread regions of neocortex and thus could potentially participate in inhibiting (or activating) engrams. Such a pathway might be the mediator of Edward Walker's (personal communication) old idea that alternation is due to an inhibition of a cortical engram corresponding to the just-entered alley. It is possible to construct a scenario in which any of these pathways (or all at once) could mediate SAB. It would be desirable, however, for the scenario to also account for the participation of the hippocampus in human memory. The following is my speculation on that topic, focusing on only a few of the possibilities described herein.

*An Explicit Model of SAB and Memory*

To develop a model in which the hippocampus simultaneously diverts attention and is involved in human conscious memory I will use the following facts and observations. First, many hippocampal neurons fire specifically to a given spatial location (see O'Keefe & Speakman, 1987). In some cases the neurons appear to be firing to a pattern of distal cues (a "frame"), but other neurons fire to a given place even when distal cues are not available (Hill & Best, 1981). The latter become unattached to the place when the rat is disoriented by twirling, suggesting that the hippocampus receives movement related feedback. Second, tetanic stimulation of the major hippocampal input (temporoammonic tract) or of several pathways within the hippocampus results in very long-term potentiation or memory (see McNaughton & Morris, 1987). Third, this memory is not equivalent to conscious human memory, since if conscious memories were stored in the hippocampus they would be lost after hippocampal damage (but they are not). Fourth, probably the most common mental response to hippocampal stimulation in man is déjà vu, or a false feeling of familiarity (Gloor, Olivier, Quesney, Andermann, & Horowitz, 1982; Halgren, Walter, Cherlow, & Crandall, 1978). Fifth, even when amnesic patients perform a memory task correctly (e.g., via cued recall) they still insist that they really cannot remember the items and that they do not experience feelings of familiarity when making a correct response (Elizabeth Warrington, personal communication). The simplified model in Figure 5.5 displays possible connections that can explain these observations.

The purpose of the circuit in Figure 5.5, aside from what has been discussed, is to allow an animal to ignore unimportant or unchanging stimuli without having to pay attention to them (except in the initial instance). We will walk through the system one step at a time. First, a cortical engram is activated via classical sensory input plus collateral activation of nucleus basalis. The latter results in attention to the stimulus and the projection of that engram into the hippocampus. Turning to the hippocampus, the arrival of the engram coupled with a cholinergic attention signal results in learning. Fantie and Goddard (1982) have shown stimulation of the medial septum to augment or enhance the population spike produced by stimulating the temporoammonic tract input to the hippocampus, which supports the preceding postulate. As a result of this hippocampal learning, the cortical engram can now by itself, without an attention signal, activate a hippocampal output. The output of the hippocampal version of the engram does many things. First, one branch descends to inhibit feelings of strangeness (and/or activate feelings of familiarity). Thus, you can now walk down a familiar street and feel that everything is familiar without having to first examine all stimuli and make a conscious decision as to familiarity or strangeness. If a building has been torn down, however, this will produce a new engram and attract attention.

A second branch of hippocampal output goes to the striatum and inhibits an orienting reaction to the stimulus. Note that this path is itself inhibited by the amygdala. This disinhibition occurs under many circumstances such as in a state of fear or when the animal is consistently rewarded for going to one place, a point to be explained later. A third pathway (to cortex) is also shown, but I am currently unclear about what it might do. The hippocampus does appear to have reciprocal interconnections with cortex. For example, Iwai and Yukie (1988) have traced a direct path from hippocampal sector CA1 to inferotemporal cortex, a region related to visual object discrimination and memory. One possibility is that whenever a potentially significant engram is stored in the hippocampus it retains for several hours the ability to reactivate the cortical version of that engram should a second important event occur. This allows two events separated by a long time interval to fuse or become associated. An obvious example is a rat eating poisoned food and getting sick hours later. Normal rats put two and two together and develop an aversion for the food, but rats with hippocampal lesions do not (Miller, Nonneman, Kelly, Neisewander, & Isaac, 1986). The existence of reciprocal connections with neocortex means that the hippocampus might contain an index of the memory, as has been postulated by Teyler and DiScenna (1986). Thus, rather than memories per se, the hippocampus could be said to store memories of memories. That may be why the famous patient H.M. can be successfully trained and yet not remember having been trained. Or why we can remember that we remember something without being able to remember what it is that we remember. Finally, there remains the possibility suggested by Walker that while a hippocampal engram is being held, the hippocampus inhibits inputs to the neocortical engram. This would allow an otherwise continuous stream of activity to become "frozen" into a frame or episode of memory, which would be protected from interference during consolidation. The figure shows the hippocampus blocking access to the engram (freezing it) while activating it.

We will now progress through a typical alternation test in a T maze. We assume that the rat is operating in the curiosity mode, but the animal could also be hungry provided that it does not have high-probability knowledge of reinforcement contingencies (to be discussed later). As the rat approaches the choice point on the first trial, two engrams are activated; we will call them "right" and "left." Note that both activate "strange," which is the default condition and which prevents the amygdala from blocking hippocampal action. Since both alleys or directions are roughly equal in excitatory strength, the rat wants to visit both, but chooses "right" during a momentary surge in its attractiveness. When that choice is made the cortical engram "right" is shunted to the hippocampus, where a neuron subjected to input from cortex and a cholinergic attention signal learns. That is, this educated neuron will now fire to cortical (temporoammonic) input alone. The educated neuron inhibits strangeness, inhibits the striatum from responding to the "right" engram, and freezes the latter.

On the second trial the striatal orienting mechanism is inhibited from responding to "right," the engram "right" is unavailable, and "right" is now familiar and thus does not stimulate curiosity. The engram "left" then captures the striatum and directs exploratory behavior towards it. Now let us examine some complications in the amygdala circuitry. It should be noted that the word "amygdala" has been chosen as an educated guess and that the model does not require that these functions actually occur in this structure. Note that the state "fear" both inhibits "curiosity" and disinhibits the hippocampus so that the rat can go endlessly to the engram "right." Note that hippocampal inhibition can also be blocked by a combination of the state "hunger" plus "prob. food." What this signifies is that if the rat is hungry and if the rat has almost certain knowledge that "right" has food and "left" does not, then the rat is free to endlessly visit "right." It should be noted that both aspects of knowledge are equally important, and that alternation can still occur if the rat is merely reinforced for "right," but not nonreinforced for "left."

One final point warrants discussion. Explaining two-trial alternation is like shooting fish in a barrel, but explaining alternation when the animal is given a long series of trials is quite another story. One possible explanation is in the way that the navigational system divides up the world. In real life the world is open-ended, but in the laboratory we break it up into pieces: Two pieces in a T maze and eight in a typical radial maze. Since the animal would be immobilized if the whole world were inhibited, the systematic exposure to all portions of the world must result in a selective disinhibition of the earlier elements in the chain. Thus, the east-turning rat inhibits east, and when it completes the circuit by going west, then east is disinhibited. In a radial maze disinhibition starts only after the eighth alley is entered and should apply primarily to the earliest-visited alleys rather than to all alleys in general. A test of this idea would be fairly easy to carry out.

I believe that the amygdala is a good candidate as a disinhibitor because of its convenient input into hippocampus-related striatal regions and the fact that it is by far the most emotional portion of the telencephalon and that its removal makes animals both tame and insensitive to reinforcement (Douglas & Pribram, 1966). There is also some evidence that in at least one case the amygdala appears to function as a disinhibitor. That is, the amygdala powerfully projects to the presumably inhibitory ventromedial nucleus, and yet lesions of the two regions have precisely opposite effects on emotionality (Albert & Walsh, 1984), which suggests that the amygdala is inhibiting an inhibitor.

## Conclusions

A staggering amount of evidence links spontaneous alternation to two sets of brain structures. One of these is the hippocampal system and the other is a set of vestibular-related structures. SAB appears to be completely and permanently abolished by large bilateral lesions of the hippocampus in general or of critical subdivisions of the hippocampal system. These include entorhinal cortex, dentate gyrus, hippocampus proper, septal region, and fornix. Unilateral lesions have no permanent effect on alternation, and small bilateral lesions tend to produce alternation deficits that are partial, disappear with postoperative recovery, or both. The cholinergic input is vital for hippocampal functioning (and SAB), as is a non-serotonergic input from the medial raphe nucleus. Serotonin and catecholamines may play minor roles in hippocampal functioning, but present evidence suggests that they are probably not importantly involved in SAB. On the output side, the evidence strongly suggests that the postcommissural fornix output from the hippocampus does not mediate either alternation in rats or memory in man. Instead, SAB appears to depend on the precommissural fornix and pathways descending in the medial forebrain bundle and/or a more dorsal loop that involves the habenula. The latter is somewhat shocking and will require further verification. Considerable evidence supports the view that any hippocampus-related physiological procedure that abolishes spontaneous alternation in animals also abolishes recent memory in man or vice versa.

Vestibular-related structures involved in SAB include the labyrinth, vestibular nuclei, cerebellum, and corpus striatum. The effects of lesions of the vestibular cortical projection region have yet to be investigated. A large body of evidence indicates that this set of structures is critically involved in navigation and that the lesions result in a loss of what could be termed a "spatial cognitive map." A model is presented in which the hippocampal system produces SAB by acting upon the spatial orientation system. The model also suggests how these systems might be involved in human memory.

*Acknowledgments.* This chapter is dedicated to my wife Dorothy, whose gift of a Macintosh II made it possible to tackle this formidable job; with special thanks to Bob Holson, whose unpublished paper rekindled my interest in SAB, and to whomever wrote the programs for Microsoft's "Word" and Cricket's "Draw" and "Graph."

# 6
# Neurochemical and Neuroanatomical Correlates of Behavioral Habituation and Sensitization: An Overview and Elaboration of Animal Experimentation

LARRY KOKKINIDIS
UNIVERSITY OF SASKATCHEWAN

## Introduction

A substantial body of evidence has accumulated suggesting that the waning of response strength to the repeated presentation of particular stimuli is an important aspect of the organism's behavioral repertoire. Much of this work has not only been instrumental in depicting the behavioral and physiological mechanisms subserving habituation, but has also engendered renewed interest in the role of habituation as a model for learning. While the early research of biologists and ethologists evaluated comprehensively the various parameters which influence habituation (e.g., role of stimulus and organismic factors), more recent research has concentrated on the adaptive value habituation plays in the animal's ability to meet environmental demands. For instance field work by Hinde (1954a, 1954b, 1961), which dealt with the mitigation of response strength to various predator models, elegantly demonstrated the importance of habituation in a naturalistic setting. Contemporary interest in behavioral habituation increased with the demonstration that habituation is indeed an important form of learning (Thorpe, 1963) and in some instances may involve principles similar to those governing classical conditioning (Stein, 1966).

In order to define the neurophysiological and neurochemical substrates of habituation, a number of investigators evaluated the plasticity of the nervous system upon repeated sensory stimulation (Groves & Thompson, 1970, 1973; Thompson, Groves, Teyler & Roemer, 1973). In fact, one of the most important influences in the proliferation of research in this field has come from work carried out at the physiological level. As with the behavioral parameters, the physiological correlates of habituation are ubiquitous in nature. Since habituation occurs as readily in single neurons (Parker, Miller & Groves, 1974) as it does in simple organisms (Lahue, Kokkinidis & Corning, 1975), several researchers have attempted to relate the laws governing these effects to those behaviors elicited or emitted from organisms higher on the phylogenetic scale (Groves & Lynch, 1972; Groves, Wilson & Boyle, 1974).

The literature involving behavioral habituation, as well as its physiological and neurochemical correlates, is exceptionally broad and a comprehensive review is not within the scope of this chapter. Rather, a brief description of various tasks used to examine pharmacological effects on behavioral habituation will be presented, followed by a review of the role of four major neurotransmitter systems–serotonin (5-HT), dopamine (DA), norepinephrine (NE) and acetylcholine (ACh)–in modulating habituation processes.

In examining this literature I will attempt to clarify and dissociate effects on habituation from the emergence of response biases which may be introduced by pharmacological and physiological manipulations used to determine the anatomical and neurochemical correlates of habituation. In particular, owing to the lack of specificity of many surgical techniques and pharmacological agents, several interpretational problems arise. For example, it is well established that d-amphetamine typically disrupts spontaneous alternation behavior (Adkins, Packwood & Marshall, 1969), a task thought to reflect habituation of exploratory tendencies (Carlton, 1969; Dember & Earl, 1957; Dember & Fowler, 1958; Douglas, 1975). This observation may be taken to indicate that the catecholamines are involved in habituation. Alternatively, amphetamine administration elicits a variety of behavioral effects including circling (Jerussi & Glick, 1974; Kokkinidis, 1987; Kokkinidis & Anisman, 1977), stereotypy (Fog, 1969; Randrup & Munkvad, 1970), locomotor excitation (Anisman & Kokkinidis, 1975; Carlsson, 1970), as well as stimulus perseveration (Kokkinidis & Anisman, 1976a, 1976b). Thus, the possibility must be entertained that the effects of amphetamine on spontaneous alternation behavior (SAB) might be an artifact of behavioral changes not directly related to response inhibition.

Although habituation is a ubiquitous phenomenon, it may at this point be safe to conclude that habituation exists at a variety of levels and may be differentially influenced by stimulus parameters depending upon the species and task employed. For example, Hinde (1954b) found that habituation of the chaffinch mobbing response to owls proceeds more quickly when the trials are distributed rather than massed. In contrast, the startle response in rodents habituates more rapidly with massed stimulus presentations (Worden, 1973). From the pharmacological vantage, anticholinergics reliably attenuate habituation in several behavioral situations (Bignami, 1976; Carlton, 1963, 1968, 1969). Similar manipulations, however, are without effect on the habituation of the startle reflex (Hughes, 1984). It is apparent from data of this sort that the physiological, anatomical and neurochemical correlates subserving habituation are not always congruent and vary depending upon the level of behavioral analysis.

One possibility which may account for these diverse observations is that habituation is not a singular phenomenon. Groves and Thompson (1970, 1973) observed that the course of habituation involves two discrete processes. The first entails a response decrement (habituation) that develops after repeated stimulus presentation, while the second involves a response enhancement (sensitization), which becomes apparent only following strong stimulation. Sensitization is observed fairly early in training, reaching a peak and then decaying with continued stimulation. In contrast, the decremental process involves only a decreasing exponential function which reaches a nadir after many repeated stimulus exposures. The actual habituation curve represents the additive or interactive effects of these processes (Groves & Thompson, 1973; Thompson et al., 1973).

Evidence that habituation involves the interaction of a response sensitization and decrement as a function of repeated stimulus presentation poses a number of questions concerning the nature of neurochemical manipulations on habituation. As will be seen in the ensuing sections, many of the tasks used to examine behavioral habituation do not dissociate between these two processes, and problems of interpretation often arise in many pharmacological experiments.

An additional purpose of this paper is to provide a model based on neurotransmitter balances that modulate the expression of behavioral habituation and sensitization. Within this framework several factors that influence habituation will be identified, including response type, reinforcement, behavioral arousal, response inhibition, and exploration. As previously discussed, the behavioral changes observed in habituation paradigms may ultimately reflect the interactive effects of independent processes, and the discussion of neurotransmitter balances in modulating habituation will deal with these issues from a theoretical perspective.

Finally, note that the behavior of interest in this chapter is not restricted to SAB. Nevertheless, reference to pertinent SAB research will be plentiful since the SAB paradigm has been extensively employed in this context.

## Behavioral Evaluations

### *Locomotor Activity, Headpoke Exploration, and Orienting Response*

The effects of pharmacological agents on habituation of exploratory activity have been extensively evaluated. For the most part, the analysis of locomotor habituation has been confined to temporal changes in horizontal activity, with relatively few studies looking at variations in vertical activity. Even with such an analysis it must be realized that a drug may affect both these behaviors, and depending on the dose, a particular drug treatment may preferentially modify one behavior, leaving the other unaltered (Hughes, Blampied & Stewart, 1975). Several studies have evaluated the effects of pharmacological agents on the habituation of locomotor activity, but the reasons for drug-induced alterations in behavior are rarely considered. In some instances, changes in ambulation may reflect the motor activating properties of drug treatments. Alternatively, increases or decreases in motor activity may be due to drug-elicited alterations in the course of habituation.

Head dipping in a hole-board apparatus has been used as a behavioral index of habituation. Although this task has received extensive use (File, 1985; File & Wardill, 1975a, 1975b), it has a major disadvantage which is shared by the open field task. As indicated earlier, both behavioral measures may be influenced by drug-induced motor effects. Thus, a particular drug may influence habituation and concurrently elicit a locomotor bias. Consequently, as was the case for locomotor activity data, interpretation with respect to specific actions on habituation must be made with caution.

Another behavior that has been employed as a measure of habituation is the orienting reflex. Not only has this behavior received extensive use from the physiological vantage, but it has been examined from the behavioral side in order to evaluate possible relations among habituation of the orienting reflex, pseudoconditioning, and shuttle avoidance (Izquierdo & Cavalheiro, 1976). Moreover, pre-exposure to the CS (i.e., habituation of orienting reflex) results in a decrement of subsequent performance, i.e., latent inhibition (c.f., Lubow, 1973, for a review of this literature). This includes both shuttle and passive avoidance in rodents (Ackil & Mellgren, 1968; Ackil, Mellgren, Halgren & Frommer, 1969; Mellgren & Hunsicker, 1975) and reaction time experiments in human subjects (Urca & Lubow, 1975).

## *Spontaneous Alternation Behavior*

Two types of alternation tasks have typically been employed in drug studies – the standard two-trial T-maze task and a free-running Y-maze task, in which animals are placed in the maze for a set period of time and the order of arm entries is recorded. Both testing situations seem to involve external cues rather than response feedback (see Chapter 2 for details). However, the type of cue (e.g., visual, odor, directional) that is actually operative in SAB may vary across species (see Chapter 3 for cross-species comparisons).

One advantage of the free-running Y-maze version of SAB is that it enables a concomitant measure of locomotor activity to be obtained by recording the frequency of arm entries. In this way, it is possible to evaluate the separate effects of a particular drug manipulation on SAB and activity levels. An additional benefit of assessing performance in the Y-maze task is that it does not involve a between-trials handling procedure as is the case in the T-maze; accordingly, alternation performance is not confounded by the stress associated with handling (c.f., Griffiths & Wahlsten, 1974 and Chapter 8).

Generally speaking, SAB exceeds chance levels and normally varies between 60–70 percent in the Y-maze task and 80–100 percent in the T-maze. If a drug disrupts habituation, then alternation should decrease, but should not fall below chance. SAB rates significantly below chance (i.e., perseveration) would lead one to suspect effects other than, or in addition to habituation. As we will see in the sections to follow, certain classes of drugs induce chance-level alternation, whereas other agents yield perseverative behavior in the SAB testing situation.

## *Carry-Over Designs*

In a habituation setting, response strength not only declines within a session (short-term habituation), but between sessions as well (long-term habituation), although typically some degree of spontaneous recovery is apparent. If a pharmacological agent disrupts the course of habituation during initial testing, then upon subsequent retesting in the undrugged state, performance should approach that of control animals. Carlton and Vogel (1965) elegantly used such a paradigm in a water-lick habituation setting. Animals were exposed to a chamber which did not contain a water bottle. Following water deprivation, subjects were reintroduced to the chamber which now contained a water bottle, and the latency to drink was used as a measure of habituation. Presumably, there exist two competing tendencies, one to drink and the other to explore the novel environment. By pre-exposing animals to the chamber, the tendency to explore should be moderated to some extent. Carlton and Vogel found that on retesting, habituated animals approached the water bottle more quickly than subjects not previously exposed to the apparatus. A major advantage of this paradigm is that the retest is carried out in the non-drug state, thus precluding the influence of motor biases produced by drugs, which, as noted earlier, make interpretation difficult when effects on habituation are analyzed in the drug state.

A similar carry-over design has been used in the assessment of the pharmacology of SAB (Kokkinidis & Anisman, 1976b). Briefly, in a Y-maze task alternation responding will decline over days (Swonger & Rech, 1972). Accordingly, if a drug disrupts alternation via effects on habituation, then subsequent performance in the non-drug state should resemble that of naive animals (high levels of alternation performance). In contrast, drugs that modify alternation without affecting habituation directly should reduce SAB on the subsequent retest–i.e., yield chance level alternation (see sections on Acetylcholine and Catecholamines).

### *Startle Reflex*

Each of the behavioral tasks described thus far has involved emitted responses. The startle paradigm involves an elicited response to the presentation of an acoustic stimulus (acoustic startle), or a strong puff of air (tactile startle). One major difference between the startle reflex and emitted responses is that with the presentation of intense stimuli changes in sensitization and habituation processes may become apparent after various neurochemical and physiological manipulations (Davis, 1980). In studies involving emitted response types, i.e., without the presentation of strong external stimulation (e.g., SAB), or in the context of very weak stimulation (e.g., orienting response), sensitization is not typically observed. The distinction between elicited and emitted responses will prove useful in identifying profiles of drug action that include alterations in sensitization and habituation.

## Pharmacological Manipulations

### *Acetylcholine*

Based on data derived from a wide variety of tasks, it is apparent that changes in cholinergic activity influence the course of habituation. Since much of the data involving cholinergic manipulations has been reviewed previously (Bignami, 1976; Carlton, 1963, 1968, 1969; Richman, Dember & Kim, 1987), only a restricted overview will be presented here. Central anticholinergic (antimuscarinic) agents have been found to disrupt habituation in an open-field exploratory situation (Green, Joyce & Summerfield, 1975; Hughes et al., 1975). However, the effects of anticholinergics on exploratory activity may reflect a drug-induced motor bias (Hughes, 1982; Walters & Block, 1969), aversive effects associated with drug treatment (Hughes et al., 1975), or might involve novel cue properties associated with drug administration (Overton, 1966; Payne & Anderson, 1967). Moreover, when several behavioral measures are employed, all of which presumably are indices of exploration, different drug-related effects emerge depending upon the dose of the anticholinegeric examined (Hughes, 1982; Hughes et al., 1975). For example, low doses of scopolamine decrease rearing and locomotor activity as well as reactivity to novelty. Higher doses reduce responding in a novel situation, whereas vertical activity remains unchanged and ambulation is markedly increased (Stewart & Blain, 1975). Thus, it seems that habituation must be considered from a broad perspective which includes several dependent variables. This point is accentuated when it is realized that drug-induced behaviors may not be mutually exclusive (Hughes et al., 1975; Stewart & Blain, 1975). For example, an increase in horizontal activity may occur at the expense of changes in vertical activity. Accordingly, whereas one measure may reflect decreased exploration, a second measure might be interpreted to indicate increased exploration.

Similar interpretive problems exist when results concerning the effects of cholinergic manipulations on hole-board exploratory behavior are evaluated. Following treatment with scopolamine or lesions to the septal nuclei, a decrease in rate and an increase in duration was observed with respect to response-dependent changes in illumination. In the absence of illumination, however, treated animals showed habituation comparable to that of controls. Based on those data, it was concluded that the habituation deficit was the result of increased responsivity to the illumination (Feigly & Hamilton, 1971). File and Wardill (1975b) suggested that the use of novel objects as the reinforcing stimuli may be advantageous in the hole-board apparatus by minimizing drug-induced response and sensory biases which might influence the habituation measure.

It is clear from the preceding discussion that factors related to response enhancement may unduly influence habituation. Thus, it is necessary to evaluate performance under conditions in which changes in motor activity play a minimal role. The free-running Y-maze SAB is a good behavioral index of habituation (Carlton, 1968, 1969; Dember & Fowler, 1958), and at the same time appears to be independent of changes in locomotor activity (Anisman, 1975). As in other forms of habituation, SAB is modified by cholinegeric manipulations. Generally, it has been found that various anticholinergics, such as scopolamine or atropine, disrupt both T- and Y-maze SAB (Anisman & Kokkinidis, 1975; Douglas & Isaacson, 1966; Drew, Miller & Baugh, 1973; Kokkinidis & Anisman, 1976a,b), whereas treatment with the anticholinesterase,physostigmine, has been shown to elevate alternation scores (Kokkinidis & Anisman, 1976a; Squire, 1969).

In the T-maze alternation task, Meyers and Domino (1964) found that scopolamine produced perseverative behavior (i.e., alternation performance which was significantly below chance). In contrast, other investigators under similar conditions failed to find a scopolamine-induced perseverative effect (Douglas & Isaacson, 1966; Leaton & Utell, 1970). Consistent with the latter reports, a series of studies assessing the effects of ACh manipulations in a free-running Y-maze situation found scopolamine to reduce SAB to chance levels only (Anisman, 1975; Anisman & Kokkinidis, 1975; Kokkinidis, 1976a; Kokkinidis & Anisman, 1980). Blocking cholinergic activity presumably attenuated habituation, thus resulting in random arm entries and chance level alternation. Failure of scopolamine to produce perseverative behavior has also been reported with other types of alternation tasks. For example, in a two-lever double alternation paradigm, scopolamine increased the number of errors in a dose-dependent fashion, but the proportion of switch to stay errors remained unaltered (Warburton, 1972). Similarly, in a discrete trial two-lever delayed spatial alternation setting, scopolamine again was observed to increase the number of errors, but failed to produce perseveration (Heise, Hrabrich, Lilie & Martin, 1975).

The deficits in alternation performance after treatment with scopolamine cannot be attributed to other types of drug-induced response biases such as heightened locomotor activity. Anisman (1975) found scopolamine to disrupt SAB in a Y-maze task, while differentially affecting locomotor activity in three strains of mice. Alternation performance apparently is not related to the motor effects of the drug since in the C57B/6J strain a deficit in alternation performance was observed, but this effect was not accompanied by a concomitant increase in locomotion. In fact, in some cases a decrease in ambulation was observed in the C57B/6J strain.

The finding that scopolamine disrupts SAB via its effects on habituation is further demonstrated by the use of a two-day behavioral testing procedure (Kokkinidis & Anisman, 1976b). Specifically, mice that initially were treated with scopolamine on Day 1 (drug-state) alternated at significantly higher levels during the retest session (24 hours later; non-drug state) than did saline-treated animals. In fact, performance of scopolamine-treated subjects during the retest session was comparable to that of naive (non-habituated) animals, suggesting that scopolamine interfered with habituation during the initial test session. It is noteworthy that the aforementioned behavioral observations did not involve peripheral drug factors or drug-induced disruption of memory consolidation processes, and the effects of scopolamine could not be attributed to drug-dissociated responding (Kokkinidis & Anisman, 1976b).

In yet another carry-over design, namely that of the waterlick habituation task, scopolamine again was observed to attenuate habituation (Anisman, Kokkinidis, Glazier & Remington, 1976; Carlton, 1969). Reiterating, in this task water-deprived animals were placed into a test chamber containing a water bottle and the latency prior to the first lick was measured. Typically, animals pre-exposed to the test chamber (water bottle removed), approached the water bottle more quickly than naive (unexposed) animals. Pre-exposure presumably resulted in the habituation of exploratory activity, which ordinarily competes with the response of approaching the water source. Consequently, the latency to drink during the subsequent retest was reduced substantially. Administration of scopolamine or atropine during the pre-exposure session resulted in performance comparable to that of naive subjects tested in the non-drug state. On the basis of these data, it was concluded that the anticholinergics mitigated the course of habituation during pre-exposure, thereby maintaining the exploratory tendency. Using a different type of carry-over design, Oliverio (1968) showed that the disruption of shuttle avoidance after prior exposure to the CS (latent inhibition) was attenuated if animals were treated with scopolamine during CS pre-exposure. Consistent with these pharmacological findings, interruption of cholinergic activity via electrolytic lesions to the septal nuclei (Weiss, Friedman & McGregor, 1974), or the hippocampus (Solomon & Moore, 1975) also eliminated the disruptive effects of prior CS exposure in the latent inhibition paradigm.

Finally, indirect evidence for cholinergic involvement in the modulation of habituation is obtained from experiments carried out with animals at various ontogenetic levels (Bronstein, 1972a). Rats during the second week of development, in which the forebrain cholinergic system has not yet fully matured (Altman & Das, 1965), failed to habituate in an open-field exploratory task (Bronstein, Neiman, Wolkoff & Levine, 1974). However, when testing occurred during the third week of postnatal development, which coincides with hippocampal maturation (Altman & Das, 1965; Campbell, Lytle & Fibiger, 1969; Douglas, 1975), within-session activity decrements were observed (Bronstein et al., 1974). The failure of immature organisms to habituate has also been found with other behaviors, including head-dipping (Feigley, Parsons, Hamilton & Spear, 1972), latent inhibition (Brennan & Barone, 1976) and SAB (Egger, Livesey & Dawson, 1973).

The finding that lesions to the septum and hippocampus produce behavior in adult rats which is comparable to that of non-lesioned immature rats (Kohler, 1976; Roberts, Dember & Brodwick, 1962; Williams, Hamilton & Carlton, 1974, 1975), including performance in a spontaneous alternation task (Clody & Carlton, 1969; Egger et al., 1973; Stevens & Cowey, 1973; White, 1974) suggests that these limbic structures play an important role in the mediation of habituation (see Chapter 5). Consistent with the view of alternation as dependent upon the development of the hippocampus, young precocial guinea pigs, in which the dentate gyrus is mature (Altman & Das, 1967), do not differ appreciably from adults in SAB (Douglas, Peterson & Douglas, 1973).

Though the data presented thus far implicate ACh in modulating habituation, there are instances when habituation is not readily modified by cholinergic manipulations. For example, whereas between-days habituation of the startle reaction is attenuated by hippocampal lesions, within-session habituation of this response is not modified by either lesions to the hippocampus or by treatment with scopolamine (Groves et al., 1974). Moreover, cholinergic agents can have subtle effects on acoustic startle amplitude (Davis, 1980), but such ACh manipulations do not influence the within-session habituation of this behavior (Hughes, 1984; Overstreet, 1977; Williams et al., 1974).

In agreement with these pharmacological findings, Williams et al. (1975) observed that while habituation of exploratory behavior was absent in immature rats, habituation of the startle response was present during this period of development. One possible account of the apparent lack of involvement of ACh in startle habituation builds on the fact that the startle reflex involves an elicited response to strong sensory stimulation. Thus, neurotransmitters other than ACh may be involved in modulating acoustic startle (Brown, 1976; Williams et al., 1975). The observation that habituation of acoustic startle is evident in immature rats (Williams et al., 1975), coupled with the fact that monoaminergic systems are developed at this ontogenetic stage (Campbell et al., 1969; Mabry & Campbell, 1974), suggests that these neurotransmitters may subserve habituation of elicited responses.

## *Serotonin*

In an attempt to assess the role of 5-HT on habituation of the acoustic startle response, a number of investigators evaluated changes in this behavior after tryptophan hydroxylase inhibition. While some reported that 5-HT depletion induced by paracholorophenylalanine (PCPA) decreased acoustic startle (Carlton & Advokat, 1973; Connor, Stolk, Barchas & Levine, 1970; Swonger, Chambers & Rech, 1970), others found no effect of PCPA on the startle reflex (Aghajanian & Sheard, 1968; Fechter, 1974a; Overstreet, 1977). Less equivocal findings were reported by Davis and Sheard (1974a), who observed that electrolytic lesions of the dorsal and medial raphe lesions enhanced startle amplitude following repeated stimulus exposure, with no significant consequence on between- and within-days habituation of this behavior. These data suggest that 5-HT is involved in modulating behavioral sensitization elicited by the presentation of strong sensory stimuli (see also Davis, 1980).

Consistent within this conclusion, administration of low doses of LSD, which inhibit the impulse flow of raphe neurons and thereby decrease the release of 5-HT (Gallager & Aghajanian, 1975), enhance startle amplitude without altering the course of habituation (Davis and Sheard, 1974b). The behavioral effects of LSD treatment, in all likelihood, reflect changes in central 5-HT activity since brom-LSD, at a dose which has comparable peripheral actions to d-LSD (Cerletti & Rothlin, 1955), but which is less potent in depressing raphe activity (Aghajanian, Foote & Sheard, 1970), was without effect on startle (Davis & Sheard, 1974b). In agreement with the LSD data, administration of low doses of N,N-dimethyltryptamine (DMT), which, like LSD, decreases neuronal firing rates of the raphe neurons (Aghajanian et al., 1970), also enhances the startle reaction (Davis & Sheard, 1974c). Furthermore, p-chloroamphetamine (PCA) which, after an initial increase in 5-HT release (Sanders-Bush, Bushing & Sulser, 1975) is followed by a rapid decrease in 5-HT and 5-hydroxyindoleacetic acid (Fuller, Hines & Mills, 1965; Miller, Cox, Snodgrass & Maickel, 1970; Pletscher, Bartholini, Bruderer, Bunkard & Grey, 1964), produces a similar biphasic effect on the startle reflex (Davis and Sheard, 1976). That is, soon after drug treatment, when 5-HT release is increased, startle amplitude is inhibited. In contrast, if testing occurs during the period of 5-HT depletion, startle responding is exaggerated.

A puzzling finding concerning the role of 5-HT in startle sensitization is the observation that electrical stimulation of the raphe nuclei attenuates habituation to a startle stimulus; furthermore, this effect is antagonized by PCPA administration (Aghajanian & Sheard, 1968). An explanation of these contradictory results can be found in recent work by Davis and co-workers, which showed that 5-HT exerts both inhibitory and excitatory effects on acoustic startle depending on the neural region under consideration (Davis, 1980). Thus, whereas increased forebrain 5-HT depressed acoustic startle, increased 5-HT in the spinal cord facilitated startle responding (Davis, Astrachan & Kass, 1980). The enhanced startle activity after stimulation of caudal portions of the raphe reported by Aghajanian and Sheard (1968) might well have reflected the excitatory effects of descending spinal 5-HT projections from the raphe nucleus (c.f. Davis, 1980).

Recent research has indicated that forebrain 5-HT is involved in startle sensitization. As mentioned earlier, electrolytic lesions of the medial and dorsal raphe nuclei, which project to forebrain structures, facilitated acoustic startle (Davis & Sheard, 1974a). 5-HT depletion by 5,7-dihydroxtryptamine lesions of the ventral tegmentum had a similar effect to that observed after electrolytic lesions of the raphe, whereas infusion of 5-HT into the hippocampus depressed startle (Geyer, 1975). Taken together, these data indicate that forebrain projections arising from the dorsal raphe modulate startle sensitization; the finding that electrical stimulation of the dorsal raphe depressed acoustic startle provides further support for this conclusion (Davis, 1980).

An alternative method for producing sensitization in the acoustic startle paradigm is to prime animals with pre-exposure to high levels of background noise prior to behavioral testing. Under these conditions, a substantial increase in startle amplitude is observed (Davis, 1974, 1980). The facilitating influence of background noise on subsequent startle amplitude also involves 5-HT activity, since treatment with LSD further enhanced the pre-exposure effect (Davis & Sheard, 1974b). Along the same lines, Conner et al. (1970) observed that following habituation to a tone, presentation of a novel auditory stimulus resulted in a marked sensitization effect in PCPA-treated animals. In agreement with a role for 5-HT in sensitization processes, it was reported (Diaz, Ellison & Masuoka, 1974) that while lesions of central 5-HT terminals by 5, 6-dihydroxytryptamine were without effect on exploratory behavior, introduction of either a novel auditory or visual stimulus elicited a greater change in behavior (increased rearing) in drug-treated animals than in control subjects. In the PCPA and 5, 6-dihydroxytryptamine studies, presentation of novel stimuli should have resulted in dishabituation in both the experimental and control groups. The fact that the behavioral responsiveness was enhanced after pharmacological manipulation of 5-HT activity suggests that this neurotransmitter is involved in the sensitization of elicited responses, but does not play a major role with respect to emitted responses in the context of weak sensory stimulation. Thus, Deakin, File, Hyde and MacLeod (1979) found that 5, 7-dihydroxytryptamine lesions of the dorsal and medial raphe did not modify the habituation either of the orienting response or exploratory activity.

While forebrain 5-HT activity is important in modulating sensitization processes, the relationship between serotonergic function and sensitization is not a simple one. For example, though lesions of the medial forebrain bundle reduce central 5-HT by as much as 60 percent, startle amplitude remains unaltered (Yunger & Harvey, 1973). Moreover, treatment with 5-HTP in conjunction with an extracerebral decarboxylase inhibitor enhances startle amplitude (Fechter, 1974a). The interpretation of these latter two points, however, is complicated by the following: a) it is not clear whether the 5-HTP effect involves increases in spinal 5-HT, b) lesions to the medial forebrain bundle also substantially reduce catecholamine levels (Heller & Moore, 1968; Storm-Mathisen & Goldberg, 1974), and c) high dosages of 5-HTP produce an accumulation of 5-HT in catecholamine neurons as well as in 5-HT areas (Butcher, Engel & Fuxe, 1972). Thus, the possibility that the catecholamines may be involved in these paradoxical effects cannot be dismissed. As will be seen in the ensuing section, catecholamine activity plays an important role in modulating the startle reaction, and it is likely that the serotonergic system acts in concert with the catecholamines in subserving acoustic startle.

## *Catecholamines*

The available evidence suggests that catecholamine manipulations can modify both habituation and sensitization processes. Moreover, there is reason to believe that in situations involving elicited and emitted response types, DA is involved in behavioral sensitization, whereas NE activity plays a role in habituation. Thus, administration of d- or l-amphetamine enhances startle amplitude, without altering within-session habituation (Davis, Svensson & Aghajanian, 1975; Kokkinidis, 1986; Kokkinidis & Anisman, 1978a; Kokkinidis & MacNeill, 1982a). The finding that tyrosine hydroxylase inhibition by α-methyl-p-tyrosine (α-MPT) depressed acoustic startle (Sorenson & Davis, 1975) provides further support for the involvement of the catecholamines in this behavior. Paradoxically, reserpine administration was found to augment startle amplitude (Fechter, 1974b). It is not entirely clear why depletion of the catecholamines by α-MpT and reserpine should have such diametrically opposite results. Moreover, while both reserpine and d-amphetamine increase startle amplitude, these effects are not entirely congruent. Reserpine enhanced startle amplitude on the first tone presentation and did not produce sensitization over repeated stimulation (Fechter, 1974b). In contrast, d-amphetamine did not augment startle amplitude on the first tone presentation. However, over repeated tone presentations an increase in startle amplitude emerged, i.e., sensitization (Kokkinidis & Anisman, 1978a). Furthermore, both L-Dopa and clonidine antagonize the reserpine-induced enhancement of startle to the first tone presentation, whereas apomorphine is without effect, suggesting that NE is involved in the behavioral changes elicited by treatment with reserpine (Fechter, 1974b). However, recent studies showing that L-Dopa increased acoustic startle in non-reserpinized animals (Kokkinidis & MacNeill, 1982b), whereas clonidine markedly depressed acoustic startle (Davis, 1980), makes the findings concerning the effects of reserpine on acoustic startle difficult to interpret. It remains to be determined whether Fechter's (1974b) observations are the result of reserpine-induced changes in epinephrine activity (Davis, 1980) or of the development of DA hyperactivity related to the depletion of this transmitter.

In any event, the available data implicate DA in the sensitization of startle responding following amphetamine administration (Davis and Aghajanian, 1976; Davis et al., 1975; Kokkinidis, 1986). In particular, the observation that α-MpT antagonizes the amphetamine-induced startle sensitization (Davis et al., 1975), taken together with the finding that apomorphine elicits a sensitization effect which was blocked by haloperidol (Davis and Aghajanian, 1976), is consistent with a role for DA in this respect. The amphetamine- and apomorphine-induced sensitization of startle responding, in all likelihood, reflects the effects of these drugs on sensory processes. Specifically, in contrast to the sensory-induced acoustic startle, startle elicited by electrical stimulation of the ventral cochlear nucleus and the reticularis pontis caudalis, which involves the motor portion of the startle reaction, was decreased by amphetamine and apomorphine administration (Davis, Commissaris, Cassella, Yang, Dember & Harty, 1986). These data suggest that the effects on DA elicited by systemic amphetamine and apomorphine treatment modulate changes in sensory processing associated with the startle reflex (Davis et al., 1986). Such a role for DA is in agreement with the well-documented sensory neglect that develops after 6-OHDA depletion of central DA (Marshall, 1979) and the increased sensorimotor responding following intrastriatal DA administration (Joyce, Davis & Hartesfeldt, 1981).

In addition to DA, pharmacological evidence indicates that NE is also involved in modulating acoustic startle. In particular, dopamine-B-hydroxylase inhibition by FLA-63 blocks the facilitating effect of amphetamine on acoustic startle (Kokkinidis & Anisman, 1978a) and tactile startle (Handley & Thomas, 1979). Phenoxybenzamine, an α-adrenergic antagonist, inhibits the facilitated acoustic startle after l-amphetamine (Kehne & Sorenson, 1978), as well as the d-amphetamine-induced increase in tactile startle (Handley & Thomas, 1979). In contrast to the role of DA in modulating startle sensitization, however, neuroanatomical experiments suggest that NE may be important in subserving the habituation of acoustic startle after repeated stimulus presentation. Specifically, electrolytic and 6-OHDA lesions of the locus coeruleus reduce the amplitude of the tactile startle response and enhance the response decrement as a function of repeated stimulation (Adams & Geyer, 1981; Geyer, 1978). Consistent with the lesion studies, treatment with low doses of the α-2 receptor agonist, clonidine, which inhibits the firing rate of locus coeruleus cells (Svensson, Bunney & Aghajanian, 1975), substantially decreases both acoustic and tactile startle (Davis, Cedarbaum, Aghajanian & Gendelman, 1977; Geyer, Puerto, Menkes, Segal & Mandell, 1976), and more importantly, accelerates the habituation of startle responding as a function of repeated stimulus presentation (c.f. Davis, 1980).

Likewise, administration of the α 1-adrenergic blocker prazosin was observed to facilitate the habituation of this behavior (Leaton & Cassella, 1984). It is possible that descending spinal NE projections modulate startle habituation since intrathecal clonidine administration increase the decrement in startle responding after repeated sensory stimulation (Davis & Astrachan, 1981).

Taken together, the acoustic startle data suggest that DA is an important substrate for the sensitization of elicited responses, whereas norepinephrine may be involved in modulating the response decrement after repeated sensory stimulation. Catecholamine manipulations also influence emitted responses in exploratory situations, and consistent with the literature involving elicited response types, the available data indicate a differential role for DA and NE activity, as well. Thus, treatment with d- and l-amphetamine induces perseverative patterns of responding in SAB tasks (Adkins et al., 1969; Anisman & Kokkinidis, 1975; Kokkinidis & Anisman, 1978b, 1980, 1981). Unlike scopolamine, which reduces alternation to chance levels, d-amphetamine decreases alternation to levels well below chance (Kokkinidis, 1987). Using the carry-over paradigm described earlier, performance of animals initially tested with amphetamine was similar to that of control subjects after retesting in the non-drug state (Kokkinidis & Anisman, 1976b), indicating that perseverative behavior is not the result of a drug-induced disruption of habituation. Moreover, the effects of amphetamine on SAB cannot be attributed to the peripheral actions of the drug, or to drug-induced interference of memory consolidation processes (Kokkinidis & Anisman, 1976b). Parenthetically, perseverative behavior is not related to other amphetamine-induced behaviors, such as locomotor excitation (Kokkinidis & Anisman, 1976a), stereotypy (Kokkinidis & Anisman, 1980, 1981) and rotational behavior (Kokkinidis, 1987; Kokkinidis & Anisman, 1977).

It is noteworthy that while habituation is not directly involved in its mediation, perseveration does appear to be dependent upon stimulus factors. For example, perseverative tendencies produced by d-amphetamine were reduced substantially when the novelty value of the apparatus was increased (Kokkinidis & Anisman, 1978b). Conversely, pre-exposure to the apparatus immediately prior to drug treatment enhanced the perseverative response pattern (Kokkinidis & Anisman, 1976a). Apparently, the repetitive behavior produced by d-amphetamine reflects an approach rather than an avoidance tendency. If an avoidance response to novel situations was exacerbated by drug treatment, then prior habituation to the Y-maze should have mitigated, or at least reduced this tendency and antagonized the drug-induced perseveration. Since pre-exposure enhanced perseverative exploration, it is likely that this behavior reflects an increased responding to selective stimuli. Indeed, the behavioral changes elicited by d-amphetamine appear to be the result of drug-induced novel interoceptive cues involving DA activity (Schechter & Cook, 1975; Scheaer & Moore, 1974). Thus, in the alternation task, the ultimate response output relies upon the interaction between the tendency to perseverate, which is subserved by the novel internal stimulus complex produced by d-amphetamine, and habituation to the external contextual cues associated with the test situation (Kokkinidis & Anisman, 1978b).

Evaluation of developmental changes in SAB provide additional support for this position. Immature animals (i.e., second week postnatally) exhibit a high degree of perseveration when tested in an alternation task. Treatment with d-amphetamine augmented the perseverative behavior (Kokkinidis, Irwin & Anisman, 1979), and the perseveration observed in immature rats was modified readily by manipulation of the external stimulus complex. Increasing the novelty value of the compartments resulted in facilitated alternation performance (Egger, 1973a), and pre-exposure to the apparatus enhanced perseverative behavior (Kokkinidis et al., 1979). The similarity between the ontogeny and pharmacology of perseveration in a Y-maze situation and acoustic startle suggests that perseverative behavior may be a form of sensitization, involving exaggerated exploratory reactions to environmental stimuli. Moreover, the observation that amphetamine administration results in perseverative head dipping (Kelley, Winnock & Stinus, 1986), as well as perseverative patterns of locomotion (Schiorring, 1979) and operant behavior (Ridley, Baker & Haystead, 1981), indicates that one of the primary consequences of this stimulant is the elicitation of perseverative response sequences, the expression of which is dependent upon catecholamine activity.

The finding that inhibition of tyrosine hydroxylase and dopamine-B-hydroxylase antagonizes the perseverative effects of amphetamine in reserpinized animals suggests the involvement of NE in this behavior (Anisman & Kokkinidis, 1975; Kokkinidis & Anisman, 1980). Recent work, however, indicates that while DA plays a paramount role in subserving the perseverative effects of amphetamine, NE might be involved primarily in the expression of habituation, which can indirectly influence the perseverative exploratory patterns of behavior seen after amphetamine treatment.

As depicted in Figure 6.1, pretreatment with haloperidol reversed the effects of amphetamine and cocaine on SAB. This finding is consistent with previous reports (Katz & Shmaltz, 1981; Oades, Taghzouti, Simon & LeMoal, 1985) and demonstrates the involvement of DA receptors in modulating the amphetamine-induced perseverative behavior. Moreover, evidence implicating a primary role for DA in perseveration is derived from neurochemical lesion studies, in which it was found that depletion of brain DA and NE by 6-OHDA attenuated the perseverative consequences of amphetamine. Since protection of NE by administration of desimipramine did not alter the effectiveness of 6-OHDA in this respect, it was concluded that DA modulates the perseverative tendency elicited by amphetamine (Bruto, Beauchamp, Zacharko & Anisman, 1984).

In agreement with this position, chemical lesions of NE-containing neurons by the neurotoxin DSP4 did not reduce the drug-induced perseveration. In fact, as can be seen in Figure 6.2, perseverative behavior was augmented, not decreased, after NE depletion by DSP4. A similar enhancement of the perseverative tendency of amphetamine was evident in an 8-arm radial maze after DSP4 pretreatment (Bruto et al., 1984). There are several possible ways of accounting for the increased perseverative response to the drug after NE depletion. First, NE may exert an inhibitory influence on DA-mediated behaviors (Antelman & Caggiula,

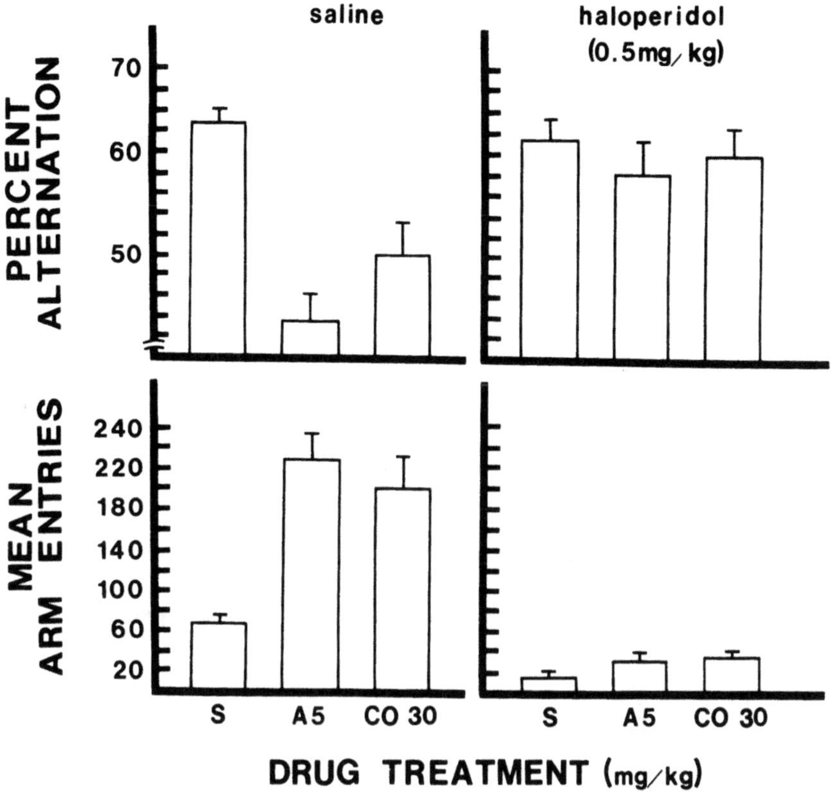


FIGURE 6.1. Mean percent (S.E.M) alternation and number of arm entries after pretreatment with saline or 0.5 mg/kg of haloperidol 1 hr prior to injection of saline, d-amphetamine (5.0 mg/kg), and cocaine (30.0 mg/kg). Behavioral testing occurred immediately after the second injection and was 15 min in duration.

1977). Thus, removal of the inhibition by interference with NE facilitates DA activity, engendering a quantifiably larger behavioral effect (Kokkinidis & Anisman, 1981). An alternative position is that DA and NE may be involved in mediating distinct processes which contribute to SAB and perseverative behavior. Specifically, while it is clear that DA subserves the perseverative effects of amphetamine (Katz & Schmaltz, 1981; Oades et al., 1985), NE appears to be directly responsible for modulating habituation, as apparently is the case for acoustic startle. The final outcome in a SAB paradigm involves the integration of these opposite behavioral tendencies. Thus, NE depletion by DSP4 facilitates the perseverative effects of amphetamine by reducing the habituation component and maximizing the drug-induced repetitive pattern of Y-maze exploration. Essentially, the behavioral consequence of NE depletion on amphetamine-induced perseveration is similar to that observed after pre-exposure to the contextual cues associated with the test situation.

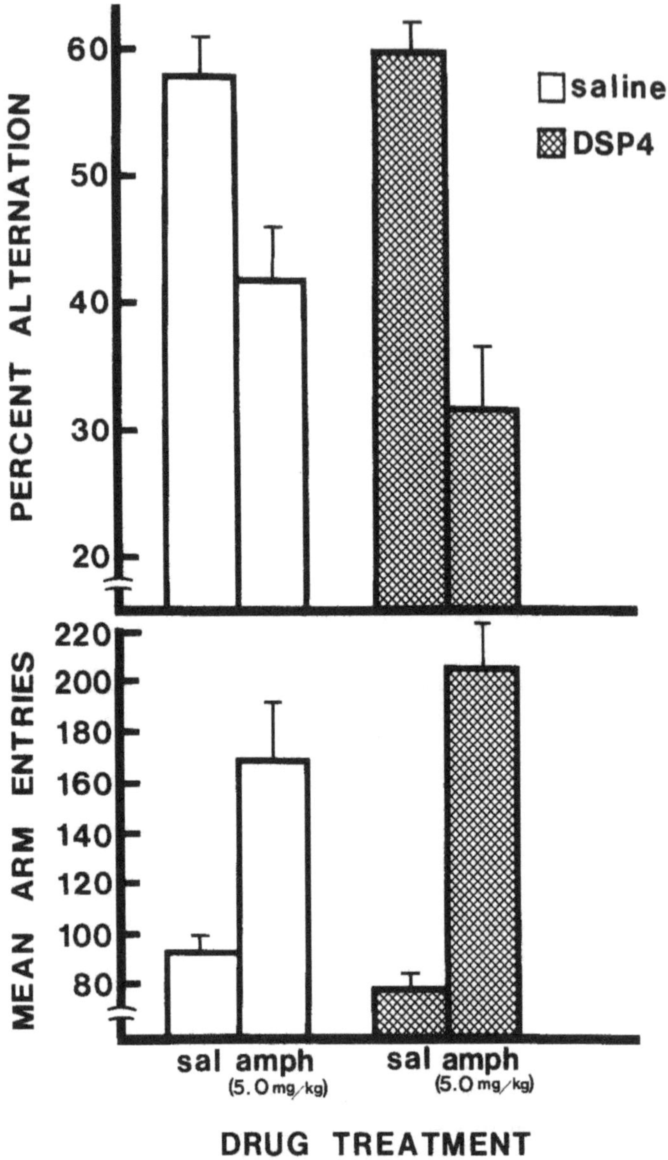


FIGURE 6.2. Mean percent (S.E.M) alternation responding and number of arm entries after injection of saline or 5.0 mg/kg of d-amphetamine in mice pretreated with either saline (white bars) or the norepinephrine neurotoxin DSP4, (hatched bars). Mice were injected with 50 mg/kg of DSP4, and behavioral testing commenced seven days later. This dose and time interval produced a substantial depletion of norepinephrine in the hypothalamus, hippocampus, cortex, and locus coeruleus (Bruto et al., 1984).

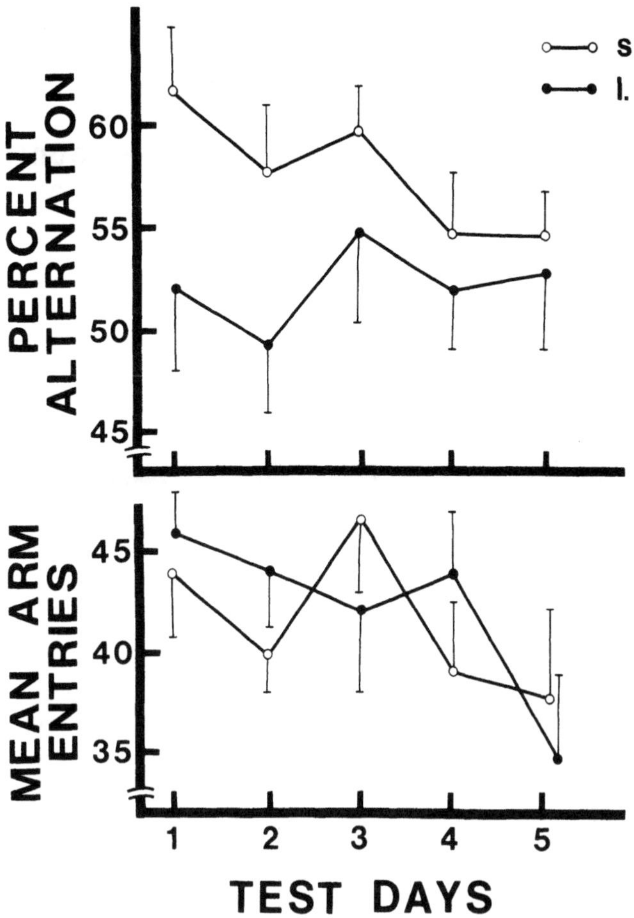


FIGURE 6.3. Mean percent (S.E.M) alternation responding and number of arm entries after bilateral electrolytic lesions of the locus coeruleus as a function of repeated daily testing. Testing was initiated seven days after either electrolytic or sham lesions, and rats were tested daily for five consecutive days.

Recent experiments involving the effects of electrolytic lesions and electrical stimulation of the locus coeruleus implicate a specific role for this brain region in subserving habituation in a SAB task. As depicted in Figure 6.3, bilateral electrolytic lesions of the locus coeruleus reduce SAB to chance levels. This observation, coupled with recent reports showing that dorsal bundle lesions after 6-OHDA treatment also decrease alternation responding (MacNaughton & Mason, 1980), indicates that NE activity is an important substrate for SAB (see also Bruto et al., 1984; Kokkinidis & Anisman, 1980, 1981). In agreement with a role for NE in modulating habituation of emitted responses, electrical stimulation of the locus coeruleus immediately prior to testing for SAB prevented the habituation of alternation responding seen as a function of repeated daily testing.

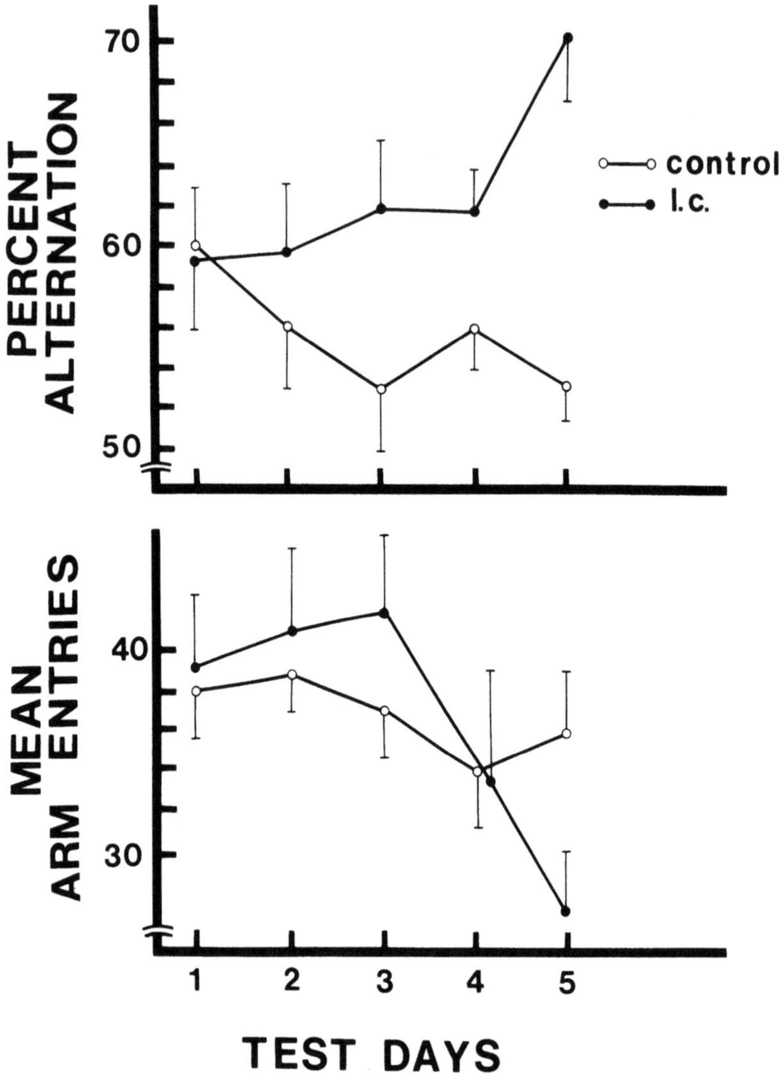


FIGURE 6.4. Mean percent (S.E.M) alternation responding and number of arm entries after bilateral electrical stimulation of the locus coeruleus prior to behavioral testing. Rats received 100 low-current electrical stimulations (10 uA, 0.1 msec pulse duration, 100 Hz, 10 sec interstimulus duration) immediately prior to testing for spontaneous alternation. This procedure was continued once daily for five consecutive days.

In fact, as is evident in Figure 6.4, five daily sessions of electrical stimulation significantly increased the strength of the alternation tendency. While the consequence of locus coeruleus stimulation on the repetitive patterns of exploratory activity elicited by amphetamine remains to be determined, it would be predicted that electrical activation of this region would have similar effects on perseverative behavior to those seen after manipulations of the stimulus novelty of the apparatus–that is, a facilitated alternation response and a reduction in the perseverative effects of amphetamine.

## Theoretical Model

The available data suggest that the course of habituation may be dictated by different neurotransmitters, depending upon the particular behavior in question. Before detailing potential neurochemical interactions involved in these processes, I will review briefly the characteristics of each transmitter with respect to its involvement in habituation/sensitization.

There is reason to believe that habituation is not affected by DA manipulations. Rather, DA stimulation modifies the habituation function via direct effects on sensitization in paradigms involving emitted and elicited response types. NE, on the other hand, appears to influence the response decrement associated with habituation of elicited and emitted responses.

The role of the cholinergic system in subserving habituation has been documented extensively. In those tasks involving emitted responses (e.g., SAB), ACh has an inhibitory role. Thus, anticholinergics attenuate the course of habituation, whereas anticholinesterases accentuate the decline in response strength otherwise observed.

Finally, the serotonergic system acts in an inhibitory capacity in the mediation of habituation; however, unlike ACh, which subserves emitted behaviors, 5-HT is involved in habituation of the elicited response type, probably by directly influencing sensitization processes.

In order to explain some of the drug interactions observed in exploratory situations, Swonger and Rech (1972, 1977) posited that the raphe-based 5-HT system acts as a gain-setter for the limbic-based inhibitory cholinergic activity. By monitoring the activity of the excitatory NE system located in the brain stem and mesencephalic reticular formation, the 5-HT system increases or decreases cholinergic inhibitory tone in order to maintain behavioral homeostasis. Thus, according to the gain-setter hypothesis, ACh is the primary source of inhibitory action, whereas 5-HT plays a secondary role, namely, that of modulating the amount of inhibition exerted by cholinergic activity in the maintenance of habituation.

While this model accounts for habituation of emitted responses, considerable difficulty is encountered in dealing with responses of the elicited variety. Recall that 5-HT manipulations directly influence this form of habituation, whereas cholinergic manipulations are without effect. Inasmuch as modification of 5-HT activity produces only a minor change in exploration, but a pronounced effect on reactivity to sensory stimulation (Brody, 1970; Davis, 1980; Deakin et al., 1979), the distinct possibility exists that elimination or reduction of forebrain 5-HT inhibition permits the manifestation of behavioral sensitization which is not otherwise apparent. Since the stimuli involved in the emitted-response type do not have the capacity to elicit a sensitization effect, 5-HT manipulations should induce little if any change in the course of habituation. Indeed, the available data indicate a lack of a 5-HT effect on exploratory behavior (Deakin et al., 1979); and in SAB, only in conjunction with catecholamine stimulants or anticholinergics do small behavioral changes emerge (Swonger and Rech, 1972).

Essentially, the position advanced in the present paper is that habituation in different situations should not be viewed in a unitary fashion. Rather, it is necessary to dissociate various portions of the habituation function on the basis of the nature and the strength of the stimuli employed, the reinforcement value of the stimuli, the response type, and the neurochemical substrates and anatomical brain regions involved in modulating the stimulus-response interface. To be more explicit, in agreement with Williams et al. (1974), habituation should be dichotomized into those types which involve emitted responses (exploration), and those in which the response is an elicited one (startle). These processes can be further distinguished on the basis that the response in the former situation may have some reinforcing properties and that the intrinsic reinforcement value associated with exploratory activity involves DA activity. Indeed, Fink and Smith (1980) recently concluded that mesolimbic DA was of major importance in the manifestation of exploratory activity (see also Taghzouti, Louilot, Herman, LeMoal & Simon, 1985).

Given the importance of DA in modulating reward processes (Fibiger, 1979; Wise, 1978), it might well be the case that exploration reflects the reinforcing impact of environmental stimuli. By way of an example, when an organism is placed into an exploratory situation, the behavioral arousal associated with the initiation of exploration is reinforcing and involves DA activity. When exploration of the novel environment is completed, and further responding does not yield reinforcement, response strength declines. The decrease in exploratory activity involves ACh activity in concert with NE from the locus coeruleus. It must be underlined at this point, however, that the effects of ACh and NE are restricted to the tuning out of irrelevant stimuli (NE) and response inhibition (ACh), without an appreciable effect on the reinforcing component of exploration. Consequently, NE and ACh manipulations are incapable of producing stimulus perseveration, a behavior which in all likelihood reflects a drug-induced exaggeration of the reinforcing value inherent in exploration. Thus, DA agonists affect behavior by exacerbating the reinforcing properties of a novel situation, without affecting the inhibitory component *per se*. Accordingly, a balance between DA (exploration) and ACh/NE (habituation) modulates the response strength of emitted behaviors (see Table 6.1).

Consistent with such a formulation, concurrent injections of scopolamine and d-amphetamine, the latter in a dose which by itself does not produce perseveration, effectively elicited perseverative exploratory patterns of responding in a Y-maze (Kokkinidis & Anisman, 1976a). Moreover, NE lesions facilitated the perseverative effects of amphetamine (Fig. 6.2). Presumably, NE depletion and scopolamine reduce the tendency to alternate, and maximize the amphetamine-induced perseveration. It follows that increasing cholinergic activity would result in decreased levels of perseverative behavior. Indeed, physostigmine facilitates

TABLE 6.1. Theoretical model concerning the neurochemical and neuroanatomical correlates of habituation and sensitization of elicited and emitted responses

| Behavior | Neurochemical | Brain Area |
|---|---|---|
| | Emitted Response (Spontaneous Alternation Behavior) | |
| exploration/ perseveration ↑ | Dopamine ↑ | mesolimbic |
| Habituation ↓ | norepinephrine (tuning out irrelevant stimuli) ↓ | Locus Coeruleus (dorsal bundle) |
| | Acetylcholine (response inhibition) ↓ | Septal-Hippocampal Axis |
| | Serotonin ↑↓ (?) | raphe (?) |
| | Elicited Response (Acoustic Startle Reaction) | |
| Arousal/Sensitization ↑ | Dopamine ↑ | (?) |
| | Serotonin ↓ | dorsal raphe (forebrain projection) |
| | Serontonin ↑ | Spinal Cord |
| Habituation ↑ | norepinephrine ↓ | Locus Coeruleus (spinal cord projection) |

SAB and attenuates the amphetamine-elicited perseveration (Kokkinidis and Anisman, 1976a).

Like the emitted type of behavior, habituation of elicited responses needs to be considered as the result of overlapping processes involving both a sensitization effect and a response decrement. In contrast to its impact on the habituation of exploratory behavior, ACh does not play a major role with respect to changes in the strength of elicited responses after repeated stimulus presentation. Rather, habituation is modulated primarily by NE, whereas behavioral sensitization involves the interaction between central DA and 5-HT systems (see Table 6.1).

As mentioned earlier, manipulation of ACh activity has no effect on the habituation of responses elicited by the presentation of strong stimuli; indeed brain stem NE seems sufficient for this form of behavioral adaptation. This conclusion is further supported by the observation that startle habituates readily after repeated sensory stimulation in young animals in which forebrain inhibitory systems have not yet matured, and from electrophysiological studies showing that the firing rate of locus coeruleus neurons to sensory stimulation habituate rapidly after repeated stimulus presentation (Foote and Bloom, 1978; Jones, Segal & Bloom, 1978). In addition, it is interesting to note that forebrain NE is not important for the habituation of elicited response types. Thus, in adult animals neither hippocampal depletion of NE, nor intrahippocampal administration of NE had any effect on tactile startle (see discussion in Adams and Geyer, 1981), suggesting that in contrast to its role vis à vis emitted responses, the development of the hippocampus is not critical for startle habituation. Moreover, these data give rise to the intriguing possibility that different NE projections arising from the locus coeruleus may modulate habituation of elicited and emitted responses.

Before closing, I want to introduce one further clarification. DA was posited to play a role in both the emitted and elicited types of responses. With respect to the former, it was suggested that DA is involved in the reinforcing properties associated with exploration of novel stimuli, whereas in the latter case this transmitter modulates sensorimotor arousal after the presentation of strong stimuli. At this time there is a paucity of data concerning the neuroanatomical correlates of the DA involvement in startle sensitization (Davis, 1980). Thus, it is not clear whether the sensitization of elicited responses also reflects the reinforcing impact of sensory stimulation, although such a possibility is certainly worthy of consideration. A summary of the proposed neurochemical balances is presented in Table 6.1.

*Acknowledgments.* This research was supported by grant A7042 from the Natural Sciences and Engineering Council of Canada.

# 7
# Ontogeny of Spontaneous Alternation Behavior

NORMAN E. SPEAR AND JAMES S. MILLER
STATE UNIVERSITY OF NEW YORK AT BINGHAMTON

Spontaneous alternation behavior (SAB) changes dramatically with postnatal development. It is impossible to discuss this relationship without paying tribute to the work of Robert J. Douglas, who has not only contributed some of the most significant studies of this relationship, but also published, almost 15 years ago, an excellent review and analysis of it in relation to the development of the hippocampus (Douglas, 1975). One feature of Douglas's review was consideration of several experiments that assessed age-related differences in SAB among human children. It is useful at this point to emphasize two lessons these studies provide us; the first of these requires substantial elaboration, but the second lesson need be described only briefly.

The first lesson is that species as diverse as humans and rats are alike in showing an increase ontogenetically in their probability of SAB. This generality is shown in Figure 7.1 (Douglas, 1975), which shows the changing rate of alternation during preschool years, and Figures 7.2–7.4, which show the ontogenetic increase in the spontaneous alternation of maze arms by the developing rat as determined from studies conducted in France (Hess & Blozovski, 1987), the United States (Bronstein, Dworkin, & Bilder, 1974), and Australia (Egger et al., 1973). It is of course uncertain that all species would show the age-related increase in SAB observed in these figures; all species have not been tested. What can be said, however, is that those species that have been tested have all shown ontogenetic increases in SAB. This includes the precocial guinea pig, which is relatively well developed and self-sufficient at birth, as well as altricial animals such as people, rats, and cats (Douglas, Peterson, & Douglas, 1973; Frederickson & Frederickson, 1979). The rate of increase in SAB with ontogeny in the guinea pig is not as rapid as that in the rat and achieves adult levels as early as 10 days of age. Yet there is so little significant postnatal maturation of the guinea pig brain that any postnatal increase in SAB is notable and probably can be attributed to experiential factors.

The details of the procedures used in the experiments (Figures 7.2–7.4) with the rat will be described shortly. More important at this point are the details of Douglas's experiment with children, because they help illustrate the second lesson.

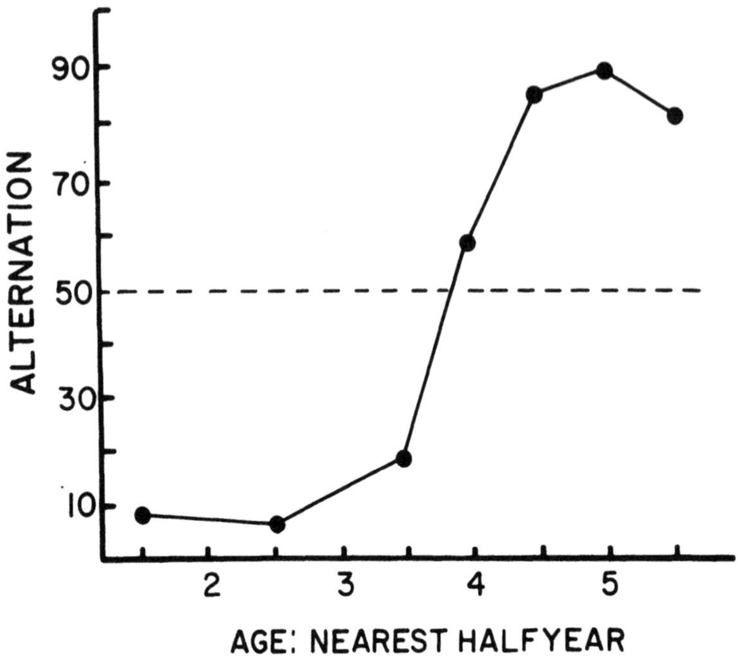


FIGURE 7.1. Percent of spontaneous alternation in children as a function of age (from Douglas, 1975).

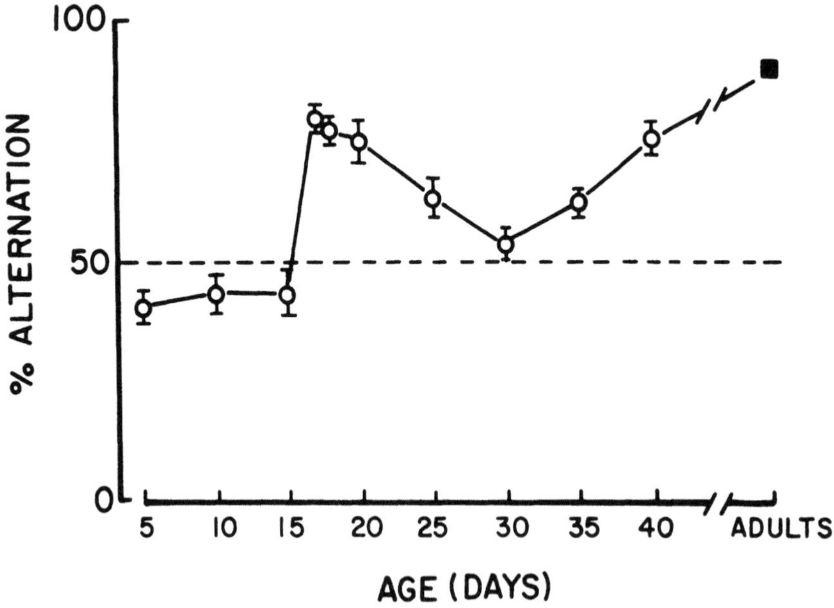


FIGURE 7.2. Percent of spontaneous alternation in rats as a function of age (from Hess & Blozovski, 1987).

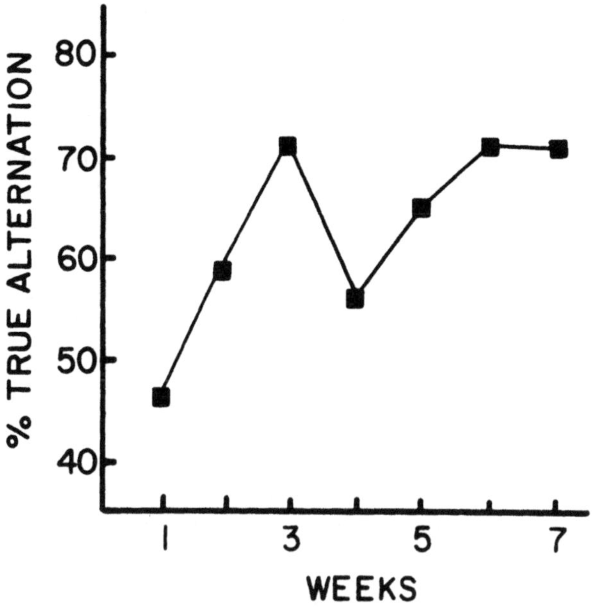


FIGURE 7.3. Percent of spontaneous alternation as a function of age in rats tested at various points during development (from Bronstein et al., 1974).

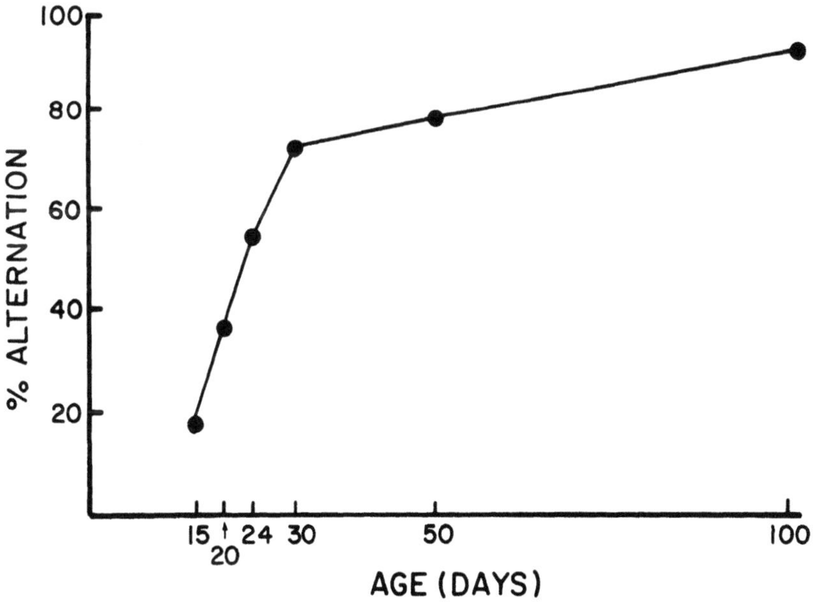


FIGURE 7.4. Percent of spontaneous alternation in rats as a function of age (from Egger et al., 1973).

On each trial Douglas allowed children to look at a picture that the child itself produced by pushing a button or pressing a lever. In front of each child were two such buttons (or levers). Pressing a button produced a picture. On each of more than 20 trials a repetition of the previously pressed button resulted in the same picture as before but an alternation – pressing the alternative button – produced a new picture. The results are best expressed in the words of Douglas (1975, p. 340): "It would be difficult to imagine clearer results. Every last individual aged 49 months or older was reliably above a chance 50% rate. . . . In contrast, only one child below 49 months alternated at a rate above 50%, a girl aged 45 months." Among the subjects tested were four individuals who did not alternate but instead perseverated in their responses; these were the four youngest subjects. Douglas concluded from these data that, "On the basis of age alone, one could predict with near certainty whether a subject would display perseveration, nonalternation (near 50%), or high-rate alternation" (p. 340).

Still younger humans, infants and toddlers, have been tested frequently for their disposition to repeat or to alternate their focus on a picture they have just seen. Perhaps the most widely used index of infant perception and memory includes the basic assumption that very young infants will be more likely to continue looking at a picture they have just seen than to look at a new picture. This is, of course, the opposite of what would be expected by a tendency for spontaneous alternation between stimuli. The assumption is so broadly verified that this effect early in life must be considered in any evaluation of the ontogeny of spontaneous alternation. Such repetition of focus on a particular picture is not universal among infants and toddlers, however. The question is whether age-related characteristics dictate when repetition or alternation will occur. A recent review of this issue has been published by Hunter and Ames (1988).

Hunter and Ames (1988) present a model and some data concerning when we might expect infants to show preference for novel stimuli and when preference for familiar stimuli might be shown instead. The value of the chapter is not that it describes a very tight fit between the model and the evidence; the value lies instead in the variables that their model identifies as important and in the insights they provide as to the complexity of the question of whether infants prefer novel or familiar stimuli.

Hunter and Ames propose that there may be no age-related differences in preference for novelty. The more important determinant of preference for novelty is the extent to which the information provided by the object or event tested is processed by the infant prior to the test. With sufficiently extensive processing of the information, we should expect a shift from preference for the item under study (i.e., the familiar item) to preference for an alternative item (i.e., novel). Younger subjects, however, may require more time to complete their information processing, and so for a given amount of study time, the younger infant might maintain a preference for the familiar item even though the older infant switches to a preference for the novel item. The other critical variable, aside from processing time and processing efficacy, is the complexity or "difficulty" of the item in question. More difficult items require more processing, and so for a given amount of study time such items may yield preference for familiarity rather than preference for novelty.

The success of this model is as yet undetermined. Appropriate tests will require independent definitions of task difficulty and effectiveness of processing, as well as effective processing time. This model, however, represents a much more advanced level of analysis than has been achieved so far in studies of the ontogeny of SAB, and it is a level to which we should aspire. The model of Hunter and Ames also provides a caveat with respect to conclusions as to the relationship between SAB and age per se, which translates something like this: It is one thing to say that the 16-day-old rat is less likely than the adult to alternate from A to B because of its undeveloped experiential history or neurophysiological equipment, and quite another to say that the 16-day-old rat is less likely to alternate because within the time allotted it has not processed A sufficiently, but would alternate at the same rate as the adult if the level of processing achieved for alternative A were equivalent.

With an eye toward the sort of analysis that might be possible and the caveat with respect to overgeneralizations about the relationship between age and preference for novelty, we may in the meantime consider some data with somewhat older human children in tests of SAB. In doing so, we should be aware that the determinants of preference for novelty in pictures and of SAB need not be completely isomorphic.

Additional studies examining SAB in human subjects have found similar relationships whether the stimuli to be alternated were pictures, simple lights (Jones, 1970), or alternative alleys in a giant T maze (Pate & Bell, 1971); and herein lies the second lesson. It is that the conventional ontogenetic increase in alternation may be expressed with any of a variety of stimuli to which the animal or human is disposed to respond at the particular age tested. This point is significant because the stimuli to which developing animals respond may change substantially during ontogeny, along with the changes that take place in the sensory and perceptual capability during ontogeny and in the ecological challenges presented the subject at any particular age (Alberts, 1981, 1984; Spear, 1984). We might therefore expect that the probability of alternation at any point in ontogeny would depend on the nature of the alternative stimuli presented. We shall return to this issue later. For now, we focus our attention exclusively on the analysis of the ontogeny of spontaneous alternation exhibited by the developing rat, the subject of the greater bulk of the analytical studies in this area.

## *Parametric Studies of Age-Related Changes in Spontaneous Alternation*

The basic features of our phenomenon may be stated simply and can be observed easily in Figures 7.2–7.4. There is a striking increase in rate of alternation between the age at which the rat's eyes first become fully open (about postnatal day 15 or 16, depending upon whether the investigator counts the date of birth as day 0 or day 1 in the animal's life) and five or so days later. Yet long after that time, for a substantial period after the rat has achieved sexual maturity (about day 50), the probability of SAB continues to increase, up to at least 100 days of age.

There is no disagreement about these fundamental facts. Many investigators from all around the world have confirmed them. Where there is disagreement is whether SAB decreases near the end of the fourth postnatal week and then increases again thereafter, as shown in Figure 7.2 (Hess & Blozovski, 1987) and Figure 7.3 (Bronstein et al., 1974). Such a decline at day 30 was not observed by Egger et al, 1973 (Figure 7.4). The agreement in the results of Bronstein et al. and Hess and Blozovski is remarkable because whereas Bronstein's test was longitudinal, repeatedly allowing the same animals a single alternation opportunity each day between ages 15 and 77 days, Hess and Blozovski tested a different set of animals at each age, allowing each animal several alternation opportunities on a single test day. The procedure of Egger et al. included only one alternation opportunity (two choices) per animal with different groups of animals tested at each age.

Why Egger et al. did not observe the decline in SAB at day 30 that was observed in the other studies is unlikely to be resolved without careful empirical research. Any number of the several procedural differences between these studies might reasonably account for the discrepancy (including a difference in strain of rat, Wistar, used by Egger et al. and Sprague-Dawley, by Bronstein et al. and Hess & Blozovski).

Yet the interesting rationale and supplementary data of Hess and Blozovski are quite persuasive that the nonmonotonic relationship between age and alternation they observed is a significant phenomenon. Hess and Blozovski suggest that there is a progressive increase in fear manifested in the experimental apparatus between postnatal days 20 and 30, and a subsequent decrease that corresponds, respectively, with the decrease and subsequent increase in SAB during this period. With number of squeaks and boluses produced by each rat serving as an index of fear, their data compellingly support this idea. There is extensive documentation in the literature and throughout this book that a frightened rat has a relatively low probability of alternation. It is therefore reasonable that emerging emotional responses might interact with the increase in the disposition to alternate over the course of ontogeny. Perhaps the specific procedures of Egger et al. (1973) somehow minimized or eliminated these emotional responses.

Our remaining discussion will, for simplicity in communication, consider the age-related change in SAB to be a monotonic increase during ontogeny. We wish to note, however, the reasonable likelihood of as-yet-untested points in ontogeny at which the probability of SAB may deviate from a monotonic increase between birth and post-puberty adulthood (also see Douglas, 1975).

## *Why does the Developing Animal Increase Its Probability of Alternation?*

We can now address the question of what it is about infants that leads to a lower rate of alternation than is seen in adults. We could of course ask the question in exactly the opposite way: What characteristics of adults make them alternate with a rate not only greater than chance but greater than infants? Although there seems to be little difference in the meaning of these two formulations, the former may leave the impression that whatever is lacking in the infants that leads them to be inadequate or less capable of adaptation than adults also leads to their lesser alternation. This impression would be incorrect. That infants adapt as well as adults to their particular ecological circumstances is shown by the fact that infants do become adults and the species continues to exist. We must in fact leave open the opposing possibility–that infants alternate less than adults because they have more of some characteristics than do adults. That said, we can now consider two levels of explanation, one environmentally or behaviorally based and the other a more reductionistic explanation rooted in brain function.

### *Environmental-Behavioral Factors*

We present in this section a sampling of plausible reasons for the ontogenetic increase in SAB, some of which previously have appeared in the literature and some of which have not.

1. Are infants less curious than adults about new stimuli and hence more apt to repeat a previous response and reexpose themselves to the same events? This seems on the face of it to be most unlikely. Although curiosity per se is not subject to easy measurement, the persistent exploratory behavior of infants documented by casual observation and systematic study (e.g., Bronstein, 1972b; Feigley et al., 1972) suggests that younger rats have more rather than less curiosity. Moreover, younger rats, particularly preweanlings, persist much longer than older ones in moving about and exploring (perhaps) an open field (e.g., Bronstein, 1972b) and in poking their heads into a hole in the wall to change the illumination (Feigley et al, 1972). Although a variety of factors other than exploration might be responsible for such age-related differences, no one as yet has convincingly discounted the possibility of greater curiosity-inspired exploration among younger animals.

2. Do infants alternate less than adults because they are generally less active than adults? Probably not. Drugs such as amphetamine and chlorpromazine drastically affect activity levels and arousal levels, at least in a certain sense, but have no influence on SAB (Anisman, 1975; Douglas, 1975; Egger, 1973b; Einon & Morgan, 1978). It is notable, however, that pentobarbitol, which can also decrease activity, decreases SAB (Douglas & Truncer, 1976).

3. Are infants more fearful of new stimuli than adults? It is almost a truism, seen throughout this book, that when the emotional state of the rat fits the

characteristics and operations consistent with the term "fear," less SAB is exhibited. We have already mentioned that Hess and Blozovski (1987) attributed the transient decrease in alternation rate between postnatal days 25 and 30 to the corresponding increase in fear measured in terms of boluses and vocalizations exhibited by the animal in the experimental situation. Is the alternation rate of the 16-day-old low relative to that of the 21-day-old because the younger rat experiences greater fear in the testing situation? Perhaps it is. The 16-day-old in isolation from its home is hyperactive, and it is conceivable that this extra scurrying around reflects fear. A variety of other responses by the 16-day-old in a strange situation, such as adrenal-pituitary activity, occur as expected if fear was prevalent. For instance, the heart rate of an infant isolated from its home increases 25 percent within a very short period, and there is an almost immediate increase of five- or six-fold in the frequency of its ultrasounding, interpreted by some as a distress call (Hofer, 1981). An odor that the infant rat experiences during a long period of isolation can in some circumstances become aversive in that the infant rat subsequently avoids the location in which that odor occurs (Smith, Kucharski, & Spear, 1985). The inference is that the isolation itself was aversive, perhaps fearful.

When infant rats have been tested for alternation, these tests have of course taken place away from the infant's home. So it is reasonable to expect that in these tests the infant rat would exhibit the same fear-like responses that it does when otherwise isolated.

One way that the infant rat "knows" that it is isolated is undoubtedly by what it smells, because olfaction guides such a very high portion of the infant rat's behaviors. We might expect, therefore, that if the odors of the home were present when the animal's behavior is formally tested, the probability of fear-like responses would decrease. Such odors can be provided, merely by taking materials from the animal's home—conventionally, wood shavings in which the rats live and make their nest in the laboratory—and placing them under and around the apparatus in which SAB is to be tested. When odors of its home are provided in this way, the 16-day-old rat substantially increases its rate of spontaneous alternation, to a level nearly equal to that of adults (Smith & Spear, 1978; Figure 7.5).

Will still younger rats show a corresponding increase in SAB in the presence of odors from the home? We have been surprised in recent experiments to learn that the answer seems to be no. In these tests, 12-day-old rats have shown no signs of increasing their rate of SAB when shavings, and hence odors, from the home are present during the test. It is not that rats of this age are insensitive to the odors of their home. If isolated and given a choice between moving toward shavings from their home and other shavings or any other location, they readily "go home"—to their home shavings. Also, rats younger than 12-days of age acquire an aversion to an odor that has been paired with long periods of isolation from their home (Smith et al., 1985).

In our recent experiments investigating SAB in the 12-day-old rat, we have found a somewhat curious result that further confirms their sensitivity to home-

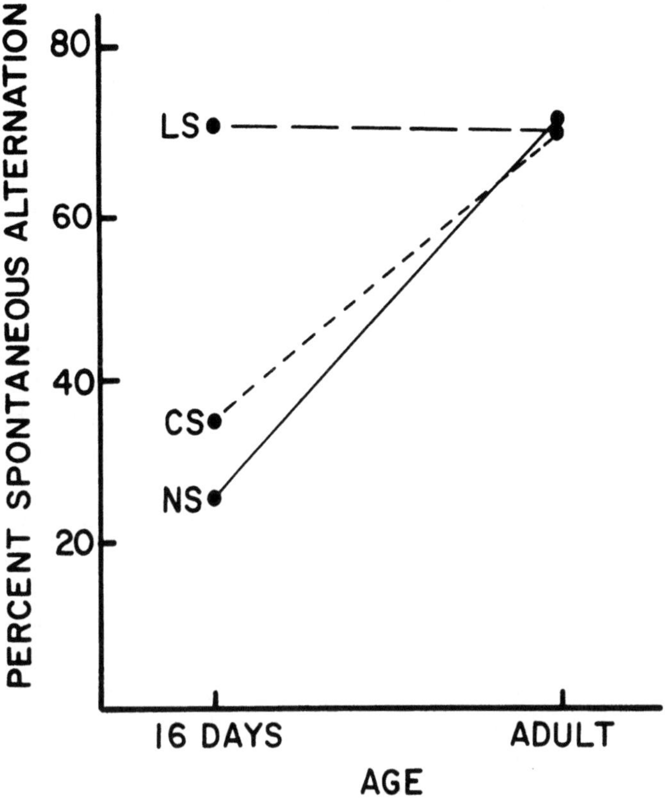


FIGURE 7.5. Percent of spontaneous alternation in preweanling (16-day-old) rats tested on no shavings (NS), clean shavings (CS), or litter shavings (LS).

like odors in terms of their SAB. Our data indicate that the 12-day-old rat does increase its rate of SAB, from the chance level of 50 percent to about 65 percent, when tested in the presence of odors from the home of 16-day-old rats. The 16-day-old rat, when tested in the presence of the odors from their home, will of course alternate spontaneously at a higher rate than otherwise. This increase in alternation rate is higher than that of the 12-day-old exposed to 16-day-old home odors, and generally replicates the effect of home odors on the 16-day-old reported by Smith and Spear (1978) using different procedures and apparatus.

It may be notable, however, that in our recent experiments alternation by the 16-day-old rat has seemed to be relatively unaffected by odors from the home of other 16-day-old rats. In view of this, the response of the 12-day-old rat to 16-day-old home odors becomes still more surprising. One might speculate that this apparently special influence of the odors from the 16-day-old's home is due to the maternal pheromone (Leon, 1974). This pheromone begins to be released by mother rats when their pups are about 14 days postnatal and so would be present for the 16-day-old's home odors but not for the 12-day-old's home odors. There is no doubt that this pheromone significantly influences the infant rat's behavior, and its effect on learning at these ages can be important. For instance, when the odors of their home are paired with a mild but punishing footshock, 12-day-old rats more readily acquire an aversion to their (nonpheromone) home odors than do 16-day-old rats to their (pheromone) home odors (Corby, Caza, & Spear, 1982).

The implications of this for the ontogeny of SAB are as yet unclear. The inclusion of home odors in the circumstances of testing could have a variety of effects other than a decrease in emotional responding that might influence spontaneous alternation, such as promoting exploration of the various odors (which could not be assumed to be equally distributed throughout the testing apparatus) or differentiation between the alternatives. When one observes the 16-day-old rat in isolation from home, probing into unknown corners, spaces, and objects with abandon and apparent fearlessness, it is difficult to appreciate that the behaviors and operations commonly used to define fear in the rat really apply to this animal.

4. Do infant rats require a longer period of stimulus exposure than adults to become familiar with that stimulus? To say that in another way, does the infant rat require a longer period for a "new" object to become "old"?

This question may be answered in terms of age-related changes in habituation. There is little doubt that habituation rate increases with ontogeny, although the particular ages covered by this relationship vary depending upon what is to be habituated. When a relatively simple stimulus is presented in such a way that stimulation of particular receptor units is held quite constant from trial to trial—as when a tactile stimulus is administered with the same physical intensity to the same spot on the animal's body from trial to trial—then there is relatively little increase in rate of habituation after the first postnatal week. In other words, when stimulation produced by the event to be habituated is relatively invariant with the animal's movement, ontogenetic change in rate of habituation does not correspond with that of SAB. On the other hand, if the event to be habituated is an object or location that can be approached by the animal from a number of different circumstances, and hence may stimulate a variety of alternative receptor sites in the sensory apparatus, then the ontogenetic increase in habituation rate is quite similar to that of SAB. For instance, Feigley et al. (1972) and Parsons, Fagan, and Spear (1973) found that between the ages of 15 and 21 days, there is a consistent and marked reduction in the rat's rate of habituation to poking its nose into a hole in the wall. As to the former case of habituation (termed "respondent" by Williams et al., 1974, as opposed to the latter, "operant," case), Campbell and Stehower (1979) observed relatively little change after postnatal day 10 in the rate of habituation to a light footshock applied through an electrode attached to the rat's foot. From this we can conclude that for some relatively complex events, at least, 16-day-old rats do require a significantly longer period of exposure than older rats to become familiar with them, that is, for the "new" objects to become "old."

Neurochemical factors that influence both the rate of SAB and the rate of habituation of the second type just described are significant here because the ontogeny of these factors is similar to that of SAB and habituation. For instance, Williams et al. (1974) found for adults that the cholinergic antagonist scopolamine decreased the rate of habituation to the nose-poking task described earlier. This drug and others with similar action have consistently been found to decrease the probability of SAB in adults (for reviews see Douglas, 1975; Kokkinidis, Chapter 6). Soon we will discuss some recent evidence by Blozovski and her colleagues that has considerably advanced our knowledge of the ontogeny of the cholinergic influence on SAB.

5. Are infants more likely to forget where they have been from one trial to the next? One surely might think so given the special disposition for rapid forgetting exhibited by infants in a variety of circumstances, an effect known as "infantile amnesia" (Campbell & Spear, 1972). This phenomenon originally referred to the more rapid forgetting observed for the events of infancy than for events of later life, but research has revealed the general fact of more rapid forgetting by infants than by adults, at least among altricial animals such as the rat or human. Among preweanling rats, rate of forgetting has seemed to be more rapid the younger the animal. The effect, therefore, may be characterized as an ontogenetic decrease in rate of forgetting, at least throughout the period of the most rapid rate of brain growth in the rat (Spear, 1978, 1979).

Tests of infantile amnesia with the rat have revealed two other of its characteristics that are particularly relevant for the ontogeny of SAB. First, the more rapid forgetting by infant rats holds for a wide variety of tasks and events to be remembered, apparently incorporating both memory for locations and responses (Campbell & Coulter, 1976). Whether SAB in infants is controlled by alternation of stimuli or responses, we may expect that the more ephemeral memory for these events in the infant rat will lead to less SAB. Second, the especially rapid forgetting by infants is observed not only over long intervals on the order of days, weeks, or months, but also over short intervals in the range of minutes and perhaps seconds (Miller, Jagielo, Spear, in press; Miller & Spear, 1989; Moye & Rudy, 1987). This is significant because it is differential forgetting over the shorter intervals, such as those used between trials in tests of SAB, that might be expected to lead to differential alternation behavior. Simply, if infants have special difficulty in remembering where they were or what they did on the previous trial in tests of SAB, we would expect their alternation behavior to be at chance level. If so, we might expect that the decrease in alternation behavior with longer intertrial intervals would be greater for infants than adults.

## *Neurophysiological Factors in the Ontogeny of Spontaneous Alternation*

One possible explanation for the reduced rate of SAB typically observed in younger animals is that brain structures important for SAB are functionally immature in the preweanling subject. For example, the emergence of SAB behavior closely parallels the development of the hippocampus (Altman & Das, 1965), and, further, bilateral lesions of the hippocampus reduce the rate of spontaneous alternation to chance levels. Interestingly, unilateral hippocampal lesions have little effect on adult spontaneous alternation, whereas the same lesions conducted at days 12–31 of age significantly disrupt alternation in adulthood (Douglas, 1975, Table 1), suggesting a more deleterious effect when hippocampal development is disrupted.

The rate of SAB also seems to be related to the level of central cholinergic activity. In adult animals, cholinergic-blocking drugs attenuate the rate of SAB, whereas pharmacological agents that enhance cholinergic activity increase SAB. The most comprehensive study examining the emergence of SAB during ontogeny as it relates to the development of the cholinergic system was conducted by Hess and Blozovski (1987). To determine the time of the onset of alternation behavior, the rate of SAB was examined in 10 separate groups of rats ranging from 5 to 40 days of age. The alternation rate for 5-day-old subjects was reliably below the 50 percent level, while 10- and 15-day-old subjects alternated at chance levels. A rapid increase in the level of SAB occurred between 15 and 17 days of age—from 40 to 80 percent—confirming the rapid onset of SAB first reported by Douglas, Peterson, and Douglas (1973).

Given this time frame for the emergence of SAB, Hess and Blozovski (1987) examined the effect of microinjections of the muscarinic cholinergic antagonists atropine and scopolamine directly into four hippocampal sites. Low doses of the antagonists disrupted SAB in rats 17 days of age or older, but not in 10- or 15-day-old rats, whereas the higher dose of the antagonists disrupted SAB at all ages tested. Hess and Blozovski suggested that although there are a number of muscarinic receptors present in the hippocampus at days 10–15, they may not become fully functional until day 17. An alternative possibility suggested by the authors is that there may be receptor subtypes that are differentially sensitive to the cholinergic antagonists. Thus, low-affinity receptors would be viewed as developing earlier and predominating in the hippocampus of the 10- or 15-day-old rat, with high-affinity receptors added by day 17. The authors do not preclude a role for the nicotinic cholinergic system or interactions between the cholinergic system and other neurotransmitters in mediating the ontogeny of SAB.

Whatever the precise mechanism is that can account for the age-dependent sensitivity to the cholinergic antagonists observed by Hess and Blozovski, it does seem clear that the emergence of SAB behavior parallels the development of the muscarinic cholinergic system in the hippocampus. It remains to be determined whether this maturation of the hippocampal cholinergic system represents an invariable physiological constraint regulating the emergence of SAB behavior, or alternatively, whether alternation will join the host of other behavioral capabilities that suddenly "emerge" when conditions are optimized by test procedures specifically designed to accommodate the unique ecological circumstances and the psychobiological dispositions of the infant rat.

### *Age-Related Changes in Stimulus Selection May Control Alternation Behavior*

Finally, we wish to consider a set of age-related factors that might contribute to ontogenetic differences in SAB, but have not previously been discussed in the literature, to our knowledge. The basic notion is that the disposition for alternation behavior in general may not change ontogenetically so much as the nature of the specific stimuli or attributes of events that control alternation. The study of the ontogeny of learning and memory advanced considerably in its progress when it was realized that the cognitive capacities of infants often are not assessed adequately by tests designed for adults (Kail & Spear, 1984; Spear & Campbell, 1979). This bit of history leads us to the following testable assertion: Whereas there is absolutely no doubt that given alternative alleys to choose and places to go in a maze, infants are less likely than adults to alternate, it is by no means certain that this same relationship with age holds for alternation between all responses and all stimuli. The events that control alternation behavior for adults may be quite different from those for infants.

The following example makes some contact with our argument. It has long been realized that performance in a T maze depends on both proximal cues within the maze and distal cues from outside the maze. Dember (Chapter 2) described the identification of the proximal, intra-maze cues (odor trails) and distal, extra-maze cues (the precise sensory basis of which is unclear, but which are undoubtedly linked to the general process of localization in space). From recent tests of spatial learning in the developing rat, we now know that prior to the age of 21–25 days postnatal, distal cues are rarely used, even in circumstances in which they are used importantly by older animals (for a review, see Spear & Rudy, in press). In all tests of the development of SAB, to our knowledge, distal cues have been available for potential use by the rat, and their use by older rats could be expected to enhance SAB for these rats but not for those younger than 21 days. It also is unclear whether infants would respond to odor trails in the same way adults do. Although the mother rat excretes a pheromone that rats can follow during their third postnatal week (Leon, 1974), there would seem to be little adaptive value for the preweanlings to respond to their own or to a sibling's odor trail during a period in which they rarely wander far from their nest.

On the other hand, the 16-day-old rat is quite sophisticated in terms of olfaction, vestibular function, and kinesthetic response. Would this animal be more likely to alternate between alternatives that are more distinctive in terms of odor, terrain, or tactile characteristics? Our preliminary tests have indicated that the presence of highly differentiable, novel odorants in alternative goal boxes does not significantly affect SAB in the 12-day-old rat. Yet because novel odorants are aversive to this animal—which might itself tend to decrease alternation—we believe that such studies do not provide a conclusive test of the idea that alternation behavior in infants and adults is governed by different events.

The general point is that the determinants of SAB may not only be fewer, in effect, for the preweanling, they may be different than those that control SAB in the adults. The overriding question is whether the age-related changes in SAB in maze behavior reflect a general ontogenetic increase in a disposition for alternation or if the relationship between age and alternation reflects instead a difference in what stimuli associated with alternative choices are most likely to control behavior at a particular age.

*Summary*

This chapter reviewed a sample of the overwhelming evidence indicating that an increase in SAB accompanies ontogeny. Consideration of the factors potentially responsible for this relationship indicated that the relatively low level of alternation by infants probably was not due to their lesser curiosity or lower general activity levels than adults. Experiential/ecological factors more likely responsible include: the fear-like responses that accompany conventional tests of alternation when infants are tested in isolation from their home; the longer period of exposure required for infants to process stimuli to a point at which a "new" alternative becomes an "old" alternative; and the more rapid forgetting by younger animals of the features of the alternative just encountered. At the level of physiological mechanisms, the hippocampus, and in particular the cholinergic muscarinic receptors, seem to mature at a rate that is correlated with the age-related changes in spontaneous alternation. Finally, it was suggested that in view of the growing evidence for age-related differences in stimulus selection in learning and memory, the events controlling SAB in infant animals are probably different from those that control it in adults.

*Acknowledgments.* This research was supported by a grant from the National Institute of Mental Health (K ROI MH35219) to Norman E. Spear. The authors wish to express their appreciation to Norman G. Richter for technical assistance and to Teri Tanenhaus and Sally Bowers for preparation of the manuscript.

# 8
# Using SAB as a Tool: Advice from a Veteran

ROBERT J. DOUGLAS
THE UNIVERSITY OF WASHINGTON

In reviewing the literature for my previous chapter in this book (Chapter 5), I was struck by the difference between the 1980s and the "heyday" of alternation studies in the 50s and 60s. First, there are now very few studies being done on this topic at all. I believe that one important reason for this is the shift to the radial maze by people who would have studied alternation in the past and who in many cases are repeating old alternation studies with a different apparatus. The radial maze measures basically the same underlying processes as alternation, but in an inefficient manner. That is, in an eight-arm radial maze it is not until the sixth trial that entering a new alley becomes improbable by chance alone. Further, it is possible for a rat to be very successful in a radial maze by adopting a circling strategy, and I have personally seen that many rats do just that. Having used both tasks, I believe that the standard SAB paradigm is more efficient and valid and that the radial maze could be more profitably employed by rewarding rats for entering a given alley rather than for avoiding it. That is, the rat would be started from a different, randomly chosen, alley on each trial and rewarded for going to a given location in space. In that way the first response would contain almost three bits of information instead of almost zero. With this procedure one would no longer be measuring curiosity but would be efficiently investigating a mechanism, spatial orientation, that is importantly involved in alternation.

My second observation was that even those few studies that do employ an alternation paradigm are using it as a tool to study something else rather than as a phenomenon to be investigated in its own right (e.g., see Lalonde's many recent studies cited in Chapter 5). In a sense, this is a sign of success rather than failure on the part of those for whom alternation is itself the object of study. In my own case, I have used SAB both ways from the outset, but I never considered the two uses to be fundamentally different. That is, as a graduate student in Robert Isaacson's lab in 1960, I simultaneously observed that my normal rats spontaneously alternated and that my hippocampally lesioned rats did not. At that point it occurred to me that if I truly understood alternation, then I would know what the hippocampus did. Or conversely, if I knew what the hippocampus did, then I would truly understand alternation. Alas, 30 years have slipped by and I still do not fully understand either alternation or the hippocampus. But I have acquired a great deal of knowledge about how to go about testing animals for SAB, and my purpose here is to pass on some of that hard-won knowledge so that you can begin where I have ended, should you so desire. I will first give some illustrations of how alternation can be employed as a dependent variable in situations where the vast majority of other tasks simply would not work. After that I will show how to deal with problems involved in alternation testing.

## *Examples of Where and When to Use SAB*

As a real-life example, some students and I decided to test the theory that thiamine deficiency was responsible for the Korsakoff syndrome. People with the Korsakoff syndrome behave as if they had hippocampal damage (see Chapter 5). Our plan, then, was to place rats on a severely thiamine-deficient diet and see if they developed behavior suggestive of hippocampal damage. We could have employed any number of hippocampus-related tasks, but nearly all of them are "one-shot deals" in which an animal is essentially "used up" after one test. This would not do, because we had no idea how long it would be before the purported hippocampal effects would occur. The solution, of course, was to employ frequent SAB tests to monitor the development of the disorder. Unfortunately, an average of 44 tests per subject over a two-month period failed to reveal any manifestation of hippocampal malfunction. We did show, however, that rats could be given about 50 alternation tests with no discernable change in rate occurring over that span.

Geoff Clark and I extended the limits when we gave literally hundreds of alternation tests to each mouse being maintained on an alcohol-containing liquid diet over a 7.5-month period and again found no trend over time. Both studies were successful in yielding unambiguous answers to our questions but were failures in the sense that editors are not enthusiastic about publishing negative results. On the other hand, the same chain of reasoning led to a happier ending when we decided to determine whether, and for how long, an electroconvulsive shock would knock out hippocampal functioning (Douglas, Pagano, et al., 1973). By using frequent alternation tests we were able to chart the entire time course of the ECS effect. In another study (Douglas, Douglas, & Treneer, 1979) we were able to plot longitudinally the rise and fall of alternation in the lifetime of the mouse. We also employed habituation for the same purpose, but found it much less satisfactory.

A related problem is one usually ignored by investigators of drug effects on behavior. In a typical drug study, for example, animals are injected and, after a wait of perhaps 15 minutes, studied behaviorally for half an hour at the most. We seldom ask how long it really takes for the drug to affect behavior or to wear off. And what if a drug has complex or even reverse effects that develop over time? Most behaviors are unsuitable for evaluating effects occurring within a very short time interval, and most tests can be given only once. Thus, it would cost a great deal in terms of money, time, and effort to blanket a reasonable time period by using different groups at each point. Alcohol and barbiturates are examples of drugs with such time-dependent properties, producing short periods of excitation (or disinhibition) followed by depressive effects. These effects can be, and have been, examined through the use of an automatic maze (to be described later) in which animals can be placed immediately after injection and studied continuously (Douglas, McGarvey, & Clark, 1984; Douglas, McGarvey, & Reilly, 1987).

Another reason for choosing SAB over, say, extinction, reversal, or passive avoidance, is that you can continue indefinitely to gather evidence from the same subjects if your first tests fail to uncover reliable effects. You can also sometimes make an educated guess as to how many tests will have to be administered to each animal in order to detect a change in rate of a given expected magnitude. As an example, Brian McDonald (1986) asked the question of whether exposure to organophosphates (e.g., insecticides) would affect memory as a consequence of the downregulation of cholinergic receptors. McDonald used passive avoidance as a task because it is generally accepted as measuring memory. Almost as an afterthought, however, he also measured SAB in these animals. It was fortunate that he did, because the variance was so high on the passive avoidance task that he failed to find significant effects despite having 20 subjects in each group and despite a large difference in group means. Since each rat was given 14 alternation tests, each could be given a fairly reliable individual score and powerful statistics could be employed. As a result, the drop in SAB scores due to organophosphate exposure was significant despite the fact that its magnitude was only 13.4 percent. Rats are not "used up" when they are given one alternation test, as some past investigators appear to have believed. On the contrary, the more tests the animal is given the more reliable the measure. Indeed, by using 25 or more tests (preferably, 50) one can begin to study the individual differences that do in fact exist among rats. One problem with alternation, however, is that it has low face validity as a measure of memory among people not familiar with this subject. This book should provide some enlightenment.

At this point, I assume that you are convinced that alternation has many unique properties as a dependent variable and that you are ready to run out, build a T maze, and test some rats. Before doing so, however, continue to read on about some of the problems that you may face. The first set of problems involves what I call "subject variables," the second, "apparatus variables," but there is considerable overlap and interaction. Over the years I have received a number of long-distance phone calls from distraught investigators insisting that they could not get their rats to alternate or, in some cases, even run in the maze. In many cases this was due to a poor choice of subjects and/or nonoptimal preparation of those animals for testing.

## *Subject Variables*

First of all, one need not automatically choose rats as subjects. Mice are preferable to rats in many ways. Apparatuses can be smaller, many more animals can be housed in the same space, and one can take advantage of the in-bred strains and numerous documented genetic defects. I have had long and intensive experience with the *Jackson Lab* strains C3H, C57, DBA, and Balb/c. The first three perform well as subjects, though even under optimal conditions they tend to alternate at lower rates than well-gentled rats. My C3H mice, for example, alternated at about the 75 percent level, and the C57s were close to 80 percent. Of the three, the DBAs have been the best all-around performers. On the other hand, do not consider using Balb/c males as subjects unless you are investigating masochism. Geoff Clark and I lost months of experimental time by buying 50 Balb/c mice which, after 20 tests each, were still not reliably above 50 percent in alternation. The reason they had so many tests was that I kept feeling that they were intelligent animals on the brink of "snapping out of it" and alternating. My hunch, incidentally, proved to be correct, but this falls into the topic of apparatus variables.

You may, however, be forced to employ rats if you intend to do stereotaxic surgery or implantation. And some people prefer a meatier subject that they can get their hands on, or one that lacks the overpowering musky odor of a mouse. One problem with rats is the great variance that can exist between and within laboratories. At least some of the commercially available rats are very poor alternators, and I strongly recommend that you do not buy a large group of them before testing a half dozen or so. Some rats are extremely maze-resistant and will not run at all. Others run but have low alternation rates. Rats called *Long-Evans* or *Sprague-Dawleys* in one lab may be quite different animals from those with similar names in other labs, so one cannot choose rats by name on the basis of the experience of other investigators at other places.

All things considered, it is best to begin with the rats in your lab. Pigmented (black-eyed) rats are preferable for a number of reasons. Hooded rats have the added virtue of being recognizable as individuals by differences in hood size and shape. If your rats have not been handled, handle them daily for about two weeks before even thinking about testing them. One of the biggest mistakes people make is attempting to save time by omitting or shortening this step. You may have to use cowhide gloves in the beginning, but the goal is to use bare hands. If you react allergically to rat scratches, you can use the glove as a perch for the rat but NOT for grabbing the animal. Semiwild rats can be picked up by firmly grabbing their tails with your bare right hand, in which case they instinctively turn away in the opposite direction and you can slip your gloved left hand under them. When you first graduate to bare hands remember to keep your fingers together, since a single digit is almost an invitation for a bite. Throughout this and later periods keep a notebook with observations about each individual. It should be noted that this whole gentling period can be reduced by rescuing rats that have been used in previous experiments and are ready to be discarded. Many departments have a course in animal experimentation, which in many cases is synonymous with "bar pressing." The students often make pets of these animals, and their training in no way interferes with alternation testing.

By the end of two weeks your rats should be distinctly friendly and accustomed to being picking up and placed down in strange places (e.g., cardboard boxes, chairs, tables, laps, etc.). They are ready to be pretested. Place the rat gently in the start box of a maze (ideas about maze construction will be discussed later) and allow it to explore at its leisure. Ten minutes is sufficient, after which the animal is returned to its home cage. You may have to repeat this treatment if the animal shows signs of freezing up. The rule is that you continue the procedure until the rat moves around confidently and unhesitatingly. On the day after this happens, you give your subjects their first two-trial alternation test. Be sure to record the latencies of all responses. On later days you can test them twice daily: once in the morning and once in the afternoon. If a rat demonstrates resistance to running, it should be replaced in the home cage and not tested again until the following day. I consider resistance to occur when a rat fails to respond within two minutes. If a rat becomes chronically resistant, discard it. Life is too short. When about 25 tests have accumulated, plot the performance of your animals across days. Your next decision will depend on what that curve displays.

If your pretraining was successful, the curve should show no trends over time. Otherwise, the curve may start at a low level (e.g., 60 percent alternation) if your animals were overly emotional, but it should plateau at a rate above 80 percent. Should your animals average below 80 percent on even the last 10 trials, then begin to examine individual data with a fine-toothed comb. The bad performance of the group could easily be attributable to a few individuals. If you identify poor alternators who show no sign of having improved, dump them and give the good rats another 10 tests or so. If your remnant group is still above 80 percent *on these additional tests* your problems are over. If not, then begin thinking about getting rats from another source. You can, however, breed good alternators from any genetically variable stock, even if you are satisfied with your group.

For breeding purposes, select the very best male, both from an alternation and a latency point of view. Breed him with several of the blue-ribbon females. The litters will probably contain very few bad alternators, provided that they are treated properly (as will be discussed later). I have spent many years trying to breed good and bad alternating rats and mice, and the results have been as follows: First, when you give 50 or more tests to a number of animals you can definitely identify individual differences. Some rats, for example, plug away at a steady rate between 60 and 70 percent while others are consistently in the high 80–low 90 percent rage. On *very* rare occasions one will find a rat that hovers around the 50 percent point.

It is easy to eliminate almost all of the bad alternators from your stock by choosing good alternators as founding parents of your lineage. In fact, most of the improvement in the stock occurs in the first or second generation born to selected good alternators. Even after six generations (nine in mice) of selection, I would still encounter an occasional bad alternator, and none of my lineages exceeded 90 percent on a day-after-day basis. In contrast to the ease of breeding good alternators, I have not been able to establish even a single lineage of poor alternators. Each of about a dozen separate lineages became sterile (pupless) and died out. On a number of occasions the mothers were seen eating their pups. Extreme emotionality was a general characteristic of these animals, which appeared to be the rodent version of the Jukes and Kallikaks. In contrast, only one of the good-alternator lineages spontaneously died out, though two of the mouse lineages picked up "bad genes" that could not be stamped out. Thus, after the first generation or two one need only delete an occasional Kallikak in order to have a stock of good subjects. Serious in-breeding can be avoided by occasionally throwing in a good alternator from another source. For reasons unknown, "bad genes" (e.g., megacolon, obese, etc.) seem to be much less of a problem in rats than in mice.

The next thing to consider is how to rear your selected animals. Although there is a large literature that is commonly interpreted to indicate that early handling is beneficial, I do not believe it is. My extensive experience includes the following observations made on hundreds of mice and rats: If rats or mice are not handled or tested until they are adults, then most of them alternate at high rates. But if you handle and test them for alternation at very early ages, then (depending on species or strain) you will produce a definite subgroup of extremely poor alternators (e.g., see Douglas, 1975). The percentage of poor alternators is higher in mice than in rats, and higher in the more emotional strains than in the more phlegmatic. Indeed, in a group of high-strung mice the majority can be permanently impaired by testing them at too early an age. This may be related to the observations of Hess and Blozovski (1987) about the emergence of emotionality during development. Their findings may explain why Douglas, Peterson, and Douglas (1973) seriously overestimated the age at which alternation develops in rats. That is, a surge of hyperemotionality occurs shortly after rats begin to develop SAB. They begin to alternate, but then fall to a low level before once again rising to an adult rate. This phenomenon can also be seen in the data of Bronstein, Dworkin, and Bilder (1974). Apparently some animals are hovering near an overemotional threshold, which is then exceeded during that emotional surge. Since I was using a criterion measure of development, this was pushed toward older ages by the temporary lapse in SAB. I have found a method that appears to get around this problem, however.

In a study not yet published, I reared random-bred mice under three conditions. One group ($n$=29, three litters) was unhandled until first tested for alternation, and a second group ($n$=21, three litters) received daily handling in which each pup was singled out from the litter and held individually for roughly two minutes. In the third group ($n$=24, three litters) the mice were touched daily as follows: Twice a day I cupped my hand over the whole litter, applying very gentle pressure, and held it there for roughly two minutes. This was possible only for about the first two weeks, after which the animals became too mobile to be so easily corralled. At that point I just put my hand in the cage. Each mouse was given one alternation test a day when between 21 and 30 days of age, and the results are shown in Figure 8.1.

It can be seen that the mice that were removed from their litters for individual daily handling performed very poorly, alternating at rates below 60 percent on both five-trial blocks. The unhandled mice performed well when 21–25 days old but then fell below 70 percent when tested between 26 and 30 days of age. Finally, the touched mice were above 80 percent on both blocks of tests. The general

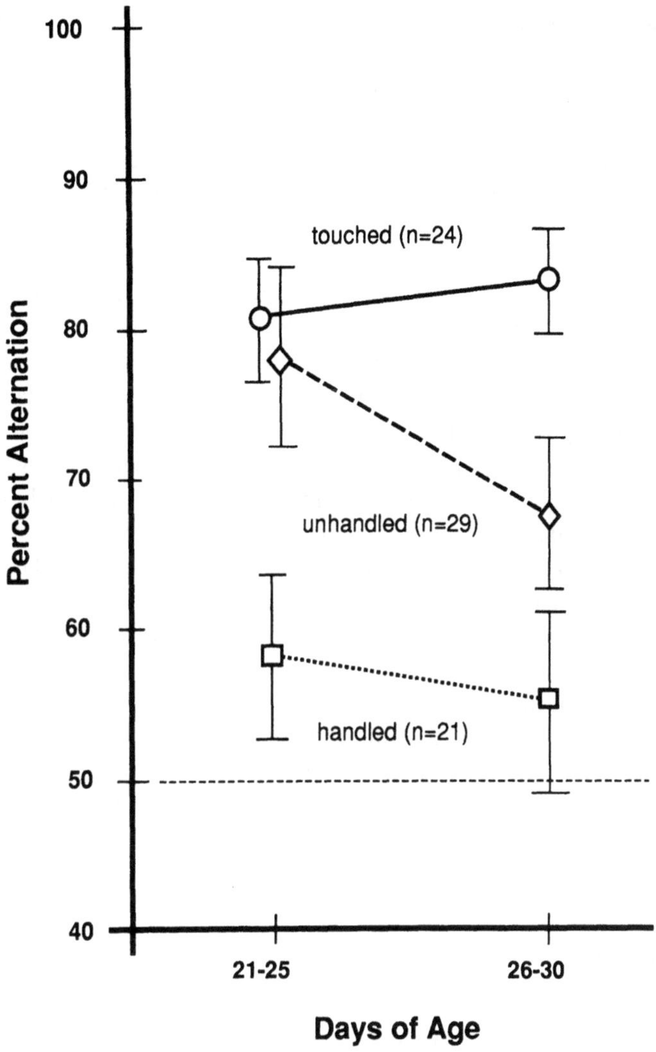


FIGURE 8.1. Spontaneous alternation on two five-test blocks in mice reared under three different conditions, as described in the text. Means and standard errors are shown. All differences are significant at the 5 percent level or better except touched versus unhandled on the first block and handled versus unhandled on the second block.

behavior of those mice at later ages, after eye-opening and the development of mobility, was also of interest. When I placed my hand in the cage they immediately jumped onto or into it, as if my hand were a magnet picking up iron filings. These mice were not merely high-rate alternators, but also extremely enthusiastic maze runners. One cannot, of course, be sure that these findings are generalizable to rats (or even to other mouse strains), but I have had such remarkably good experience with the touching technique that I strongly recommend it. I suspect that two factors are involved in its effectiveness. First, the pups may imprint on the experimenter's odor (or perhaps human smell in general). Second, by keeping the pups in the litter pile one avoids stress due to temperature drop or other factors.

You may, however, not always have the luxury of choosing good alternators as your subjects and may be forced to deal with a given group of animals that fall somewhat short of perfection. What then? First, it has been demonstrated that some poorly alternating rats have a significant improvement in behavior if sawdust from their home cages is sprinkled in the maze (Smith, Goodlett, Burright, Donovick, & Spear, 1983; Smith & Spear, 1978). If your problem is that your animals are generally sluggish and inactive, you might try the trick of using a reversed light-dark cycle. Most rodents were designed to be nocturnal and are more active during the dark phase. If this fails to work, establish each animal's body weight and determine how much food it eats per day. Then remove all food and feed each animal half of what it usually eats, giving the food only after the end of testing that day (or after the end of the intended testing time). As long as body weight remains above 85 percent of its original value there need be no concern about health. The amount of food can be gradually increased, but the initial large reduction tends to get the animals' attention. After a few days of this treatment, the animals become markedly more active. If this fails to work, keep them on their diet, but at the intended testing time put them in a cardboard box with an inch of sawdust on the floor and mix a handful of grain in the sawdust. Popcorn (unpopped) works fine. Leave them in the box until they have discovered the grain and have eaten a dozen grains or so. If after several days of this treatment you are still encountering resistance, your last recourse is to mutter something about the owls at the zoo being hungry.

## *Apparatus Variables*

From earlier chapters you have learned that many of the classical alternation studies were performed by Edward Walker and a band of dedicated students during the 50s in Mason Hall at The University of Michigan. But by the time I entered graduate school in the fall of 59, those students had scattered to different campuses and Ed Walker's magnificent eyebrows were no longer to be seen in the large lab that bore his name. All that remained were legends (known to apparently everybody except me) and a collection of well-used mazes. My mentor, Bob Isaacson, inherited the apparatus of this bygone era, and our early success in doing "hippocampal" research was based on the following formula: Take 10 rats with hippocampal lesions and test them in whatever apparatus you find lying about. I unwittingly rediscovered the phenomenon of spontaneous alternation only because Isaacson found a T maze and told me to find out which side was preferred by each rat.

In any event, I was soon doing one alternation study after another and found those inherited mazes to be a major source of aggravation. The sliding doors were too thick for the grooves they were supposed to slide in, and they often got stuck at exactly the wrong moment. The guillotine doors were made so that if you pulled slightly too hard on the string they came up out of the groove. The strings were flimsy and often broke (again, always at the wrong time). Finally, the mazes were built with floors that were virtually impossible to clean (and consequently, almost never were). Strangely, it was several years before it occurred to me to build my own mazes and eliminate these faults. Making a wooden maze (or worse, a plastic one) is not nearly as easy as you might think. Each cut has to be exactly 90 degrees or the maze will not sit flat. There is some allowable slop in the case of wood, but with plastic it is best to buy the pieces cut to size on professional equipment. Problems of sticking doors are easily solved by making wider grooves (allowing for paint) and/or using sheet metal doors. Guillotine doors can be constructed so that when you pull the string they hit a "stopper" that prevents them from rising out of the groove, and you can use high-test fishing line to reduce the possibility of a break. Problems of cleaning can be eliminated by simply not building a floor into your maze. Its floor is whatever you set it on.

My doctoral dissertation (Douglas, 1964, 1966a) involved testing 48 rats an endless number of times, and the preceding improvements made it less of a traumatic experience than it would otherwise have been. My procedure still had a defect in that once a maze was picked up it was hard to put it down in exactly the same place, especially if one used butcher paper for the immediate floor of the maze (as I did) and changed the paper between trials. This problem was later solved by making a 3.5- by 2-foot plywood floor for each maze, covering it with formica, and hinging the back of the maze to the floor such that it could be raised up (for floor cleaning—one quick swipe of a sponge—or for paper-change) and then lowered to its exact place.

But there was one final defect that was far worse than all the others put together, and that was the problem of deciding when an animal had entered an alley. Often a rat seemed to have decided upon an alley, but when I began to lower to door it started to back out. In some cases the descending door would then hit the animal on the back or the haunches, and it would jump forward into the alley. In other cases, the rat would succeed in backing out into the choice-point region. The animals appeared to be reacting to the sound of the door, and I spent a great deal of time agonizing over the possibility that I was creating the data rather than recording it. In actual fact, the rats more often disproved my ideas than proved them, but I still worried about a possible experimenter effect. The answer, of course, was to use a one-way swinging door so that all decisions would be made purely by the rat. Such doors had been used by the maze runners of the 20s and 30s but had not been employed by alternation investigators. At first I used wire mesh swinging doors, but then I discovered that some animals are bothered by the roughness and are very reluctant to have the mesh raked over their bodies as they cross through. I therefore later employed thin (1/32-in) clear plastic doors with very smooth edges, using a strip of tape as the hinge. A piece of stiff cardboard also works fine as a swinging door.

By that point I had evolved a bug-free maze; or so I thought. Its only defect was that some ungrateful animals simply refused to run in it. I discovered this problem while testing mice with genetic vestibular defects (Douglas, Clark, et al., 1979). The first mouse placed in my "supermaze" was still in the start box, as if nailed to the floor, an hour later. I placed food and water in the side alleys and left the mouse in the maze overnight. Next the morning the mouse was still in the start box all hunched up in a pool of urine and feces. Some of the other mice did run in the maze, reluctantly, but most were more like the first one. My hunch was that there was something about smooth slippery floors that the mice disliked, so I built a sort of T maze from a cardboard box and filled it about an inch deep with sawdust. Feces or urine could be removed in an instant, along with a handful of sawdust. Or the sawdust could all be dumped out (upon occasion) and replaced within seconds.

The mice all loved the cardboard maze, and I had no further problems. Because these box mazes were so easy to build, I soon had five of them placed in a row and could gather data at alarming speed by dropping one mouse after another in mazes down the line and then returning to the first maze and giving the animal therein a second trial. Often I would gather more data in a single day than were contained in any 10 "classical" alternation studies put together. Figure 8.2 shows a mouse version made from a single box and a larger rat version made from two boxes plus cardboard sheets. You can use those boxes that office paper (ditto, stationery, photostat, etc.) comes in, since many offices generate them in staggering numbers. They are about 17.5 in. long, 11.5 in. wide, and 9.5 in. tall. To prevent animals from jumping out you can cover the box with a sheet of wire mesh or, preferably, very thin clear plastic. Any large city has a company that sells (for a modest price) boxes of virtually any size, and you can buy taller boxes that require no lids. You can also buy plain pieces of corrugated cardboard to be used, along with white glue, in constructing the inner walls.

At upper right in Figure 8.2 is an automatic maze in which the animal can decide for itself how long to remain in a side alley, since one-way doors lead back from the side alleys to the starting region. Remember the Balb/c mice that did not alternate at above-chance rates after 20 tests in a standard maze? The very first one tested in the automatic maze alternated 23 out of 25 times. Their problem appears to have been a hyperreactivity to being handled. Balb/c males are so innately vicious that one tends to be afraid of them and to handle them in such a way as to exacerbate their emotionality. Emotionality is, of course, communicated from one animal to another as well between subject and experimenter. As merely one of many similar examples, I once did a study that involved weighing each animal prior to alternation testing, and one rat went beserk while being removed from the metal cage of the scale. Squealing furiously, it leaped onto the floor and managed to elude capture for about 15 minutes. When I finally caught the rat, I stupidly tried to test it in the maze. It would not leave the start box. When I then tried to test the other animals (all normally very good performers), one after another also refused to run. This might have been an example of a phenomenon that I and several others have called "maze poisoning." That is, one of your subjects may become emotional in the maze and refuse to run, and then the same will happen with the next subject, and the next, etc., as if they have

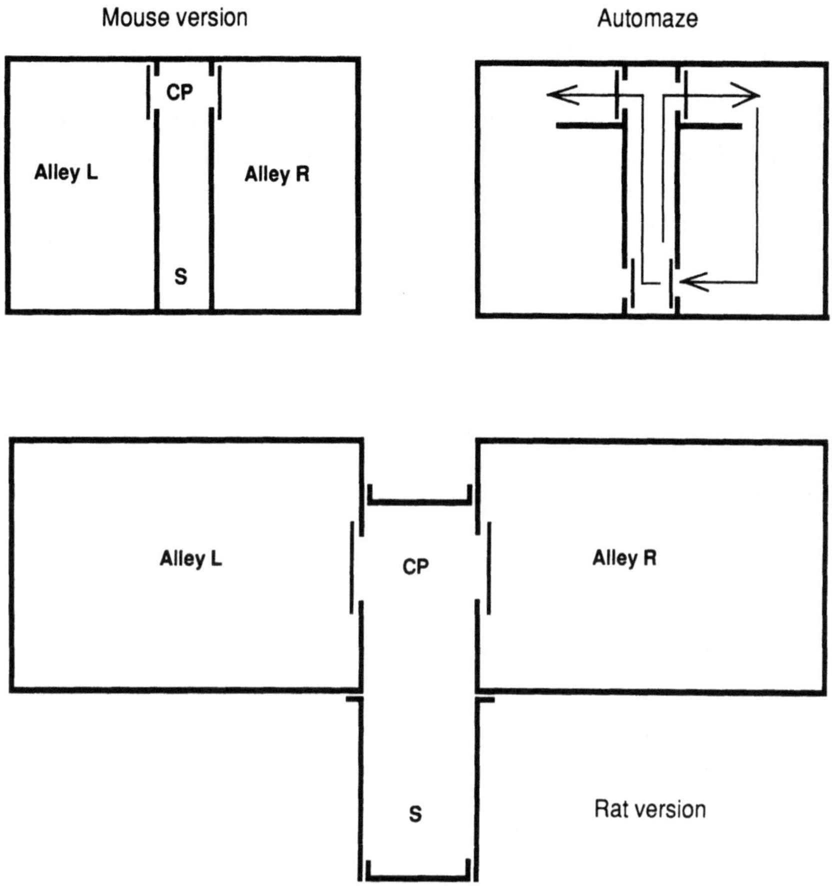


FIGURE 8.2. Examples of mazes built from standard-size cardboard boxes and scrap pieces of corrugated cardboard. Short lines indicate swinging one-way doors or, in the automaze, sliding doors.

caught some disease. A number of us have suspected that the feces or urine might contain an "alarm substance," but it is difficult to study a phenomenon that strikes out of the blue at a moment when you least expect it.

An important point about the Balb/c animals is that I would have been amply justified in claiming that males of this strain lack SAB, even though we now know that this is not true. A parallel situation is the case of rabbits, which have been claimed (and shown) to lack alternation (see Baisden et al., 1972). When I was at Claremont Graduate School, I studied a group of eight rabbits that also failed to alternate. These rabbits were fairly friendly towards me (though I did have to catch them for each test) but became extremely "uptight" in the maze. I noticed that even though the maze was large (on a rat scale), both sides of the rabbits' bodies touched the walls. I therefore built a giant maze with alleys six feet long and one foot wide. Further, I placed this maze on the grass in our patio instead of on a slippery floor inside a building. The rabbits hopped through the maze with enthusiasm and averaged almost 80 percent alternation. These examples bring up the question of the point at which one concludes that a group of animals is lacking in alternation. It is clear that any possibility of emotional reactivity must be eliminated. One rather extreme way to do this would be to make bilateral lesions of the amygdala, which eliminate emotional reactivity in a wide variety of species and have no harmful effect on alternation (see Chapter 5). Benzodiazepines (e.g., valium) in low doses might also work, though at higher dosage they abolish alternation (Iwahara, Oishi, Yamazaki, & Sakai, 1972).

The automatic maze shown in Figure 8.2 has many advantages. The most important, of course, is the reduction of experimenter-subject interactions to a bare minimum. Another is the speed with which the animals generate data. They spend far less time in the just-chosen alley than you might imagine (often only a few seconds). In an automaze it is very easy to study massed-trial alternation. Some mice will run 25 circuits within a mere 15 minutes, though half an hour is

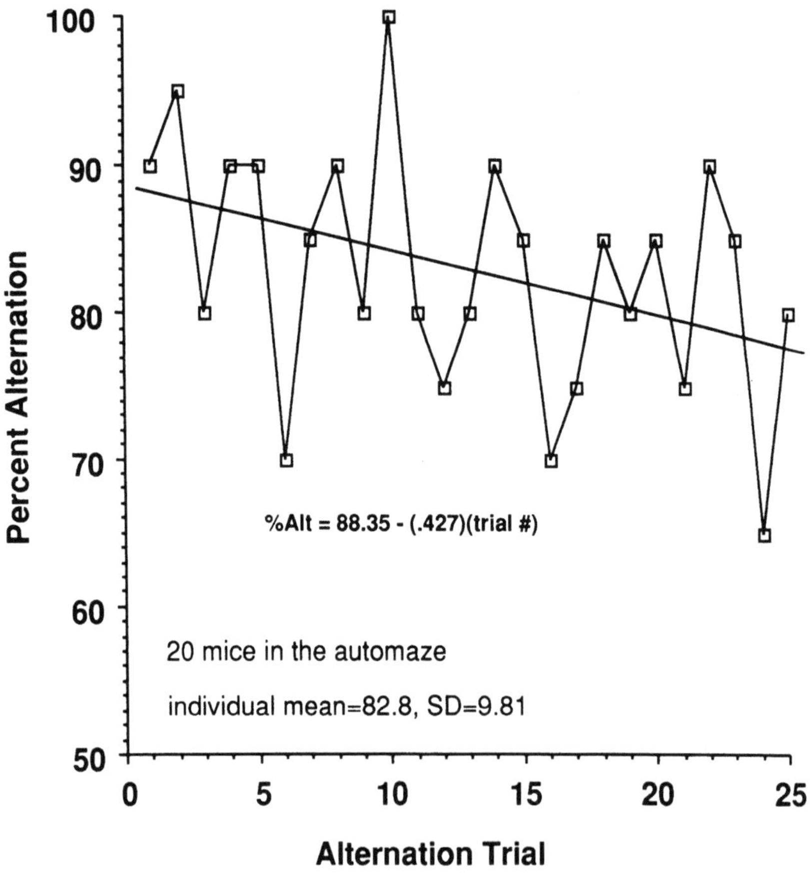


FIGURE 8.3. Spontaneous alternation across 25 massed alternation trials in 20 mice from selected stock. The equation refers to the least-squares fitted line.

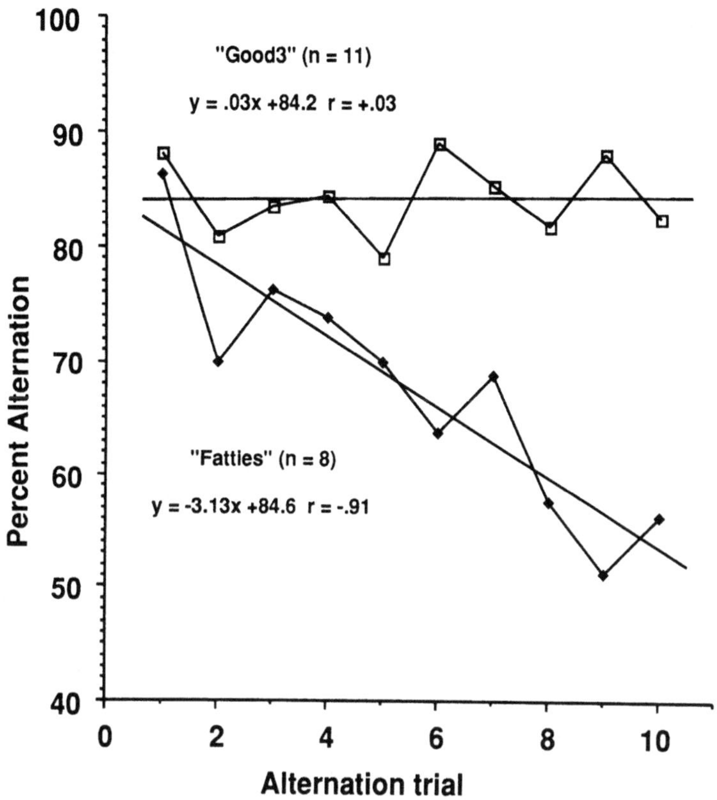


FIGURE 8.4. Spontaneous alternation across 10 massed alternation trials in mice from two litters. Each point represents the mean for 10 tests/subject.

more typical. You can use the first two trials as a measure of conventional SAB or, if there is only a slight trend over trials, use all of the data for that purpose. Alternatively, you can use all of the data as a measure of what you frankly call "massed-trial alternation." Figure 8.3 displays the results of a test in which 20 mice were allowed to respond 26 times at their leisure in an automaze. Though these mice were from a random-bred source, they were from a "good alternator" lineage that I had bred.

The sloping line is fitted by the least-squares method and suggests a gradual decline over 25 trials. The correlation between trials and alternation is actually not significant ($r = -0.38$) in this particular instance, but a decline of the magnitude shown in Figure 8.3 is typical when this many massed trials are used and is probably genuine. In this case one could claim that the alternation rate is 90 percent (first two trials), 88 percent (curve extrapolation for test 1), or 83 percent (the group mean for all tests). With data of this sort one can give each animal two independent scores that correspond to the intercept and the slope. The former is mathematically the "best bet" for two-trial alternation, and the slope clearly measures some other factor (interference?) that could be of considerable interest. Figure 8.4 shows the behavior of two different litters when each mouse was given

11 trials per session in the automaze on 10 different occasions. One litter was the third generation in a "good alternation" lineage and the other was born to a pair of obese piebald mice bought at a pet store.

It can be seen that the "Good3" lineage displays no sign at all of any decline across tests whereas the "Fatties" decline markedly. Least-squares curves were calculated for each individual, and the difference between groups in slope was highly reliable ($t = 3.8, p < 0.01$) even though the two groups were almost identical on the intercept for alternation trial 1 (84.2 versus 81.5). Of the two groups, the Good3s are very much more typical of mice in general. That is, when mice are given 11 consecutive trials in an automaze, they usually have a flat curve. In that case one is justified in using all of the data to calculate the SAB rate. When this is done for the Good3 litter, the mean is 84.4 percent, and the standard error a mere 1.78. The latter indicates that we can be certain at the 5 percent level of confidence that the alternation rate was above 80.4 percent and below 88.3 percent. This high degree of precision required only about an hour of testing per day for 10 days. If these animals had not been selected, however, the standard deviation would probably have been higher and the confidence interval more like 78–91 percent. The Fatties have the steepest decline I can ever remember seeing in normal mice, but I suspect that certain brain lesions or drugs might be found in the future to produce such an effect, in which case you will know how to deal with it.

Additional possibilities arise when a more fully automated, preferably computerized, version of the automatic maze is employed. That is, instead of swinging doors you can use doors that are raised and lowered by solenoids, with the animal's movements being monitored by photocells. Memory, for example, can be systematically studied by allowing the animal to reenter the choice region but not opening the choice doors until a predetermined period of time has elapsed. The time that the animal spends in a side alley can also be systematically varied or controlled. In the case of the two litters shown in Figure 8.4, alley exposure time was not controlled and might have contributed in an unknown way to the differences in slope. Needless to say, any number of such mazes can be used simultaneously and mountains of data gathered and analyzed by the uncomplaining computer. By these methods one can gather so much data that it becomes possible to make fine distinctions between levels of performance rather than merely (as in the old days) trying to decide whether the animals are alternating or not. By analogy, an ophthalmologist asks not whether you can see, but how well you can see.

Although I imagined myself to have invented all of the apparatus features described in this chapter, many others had made the same discoveries years or even decades ahead of me. I have recently come across at least half a dozen studies employing "my" automaze (e.g., the "self return" maze of Greene & Naranjo, 1986) or a reasonable facsimile thereof. In most cases, automatic mazes have been used in experiments involving learning and rewards rather than for testing SAB. Spontaneous alternation has been tested in automated symmetrical Y mazes in which the animal theoretically has three choices upon exiting from a given alley instead of the usual two (e.g., see Chapter 6). There is no reason why one has to stop at three alleys; about 15 years ago Denis Mitchell and I built an eight-arm version of the automaze in which the animal returned to the central arena by a path underneath the outbound alley. I personally favor the two-choice situation for a variety of reasons, including the link to the literature of the past.

In summary, by following a few simple rules the chances are very high that your animals will run with enthusiasm in mazes and alternate, should they possess the neural mechanisms required to do so. As discussed, you may encounter some special problems, but I think that you will now be able to go about systematically solving them. If you are impatient, as I am, you should keep in mind that every minute spent in preparation saves an hour of grief. Remember the keys to the use of alternation as a tool. First, acquire or generate good subjects. Second, build bug-free mazes. Third, develop a standard procedure for running and evaluating your animals. Fourth, insure reliability by testing each animal many times and giving individual alternation scores to each. As a possible fifth, you might give these tests over a broader time span and thus monitor such changes as might occur. From the latter data you may discover many things that would not show up in one short testing period. Finally, remember that the odds are with you. If there is something worthwhile to be discovered, you will probably find it.

*Acknowledgments.* This chapter is dedicated to those who made this chapter necessary: the rats and mice who discovered so many ways to make life difficult.

# References


Abraham, L., Copack, P., & Gilman, S. (1977). Brain stem pathways for vestibular projections to cerebral cortex in the cat. *Experimental Neurology, 55*, 436–448.

Abraham, L., Potegal, M., & Miller, S. (1983). Evidence for caudate nucleus involvement in an egocentric spatial task: Return from passive transport. *Physiological Psychology, 11*, 11–17.

Ackil, J.E., & Mellgren, R.L. (1968). Stimulus pre-exposure and instrumental learning. *Psychonomic Science, 11*, 339–344.

Ackil, J.E., Mellgren, R.L., Halgren, C., & Frommer, G.P. (1969). Effects of pre-exposure on avoidance learning in rats with hippocampal lesions. *Journal of Comparative and Physiological Psychology, 69*, 739–747.

Adams, L.M., & Geyer, M.A. (1981). Effects of 6-hydroxydopamine lesions of the locus coeruleus on startle in rats. *Psychopharmacology, 13*, 394–398.

Aderman, M., & Dawson, J.N. (1970). Comparison of forced-choice alternation in goldfish and planaria. *Journal of Comparative and Physiological Psychology, 71*, 29–33.

Adey, N.R., Segundo, J.P., & Livingston, R.B. (1957). Corticofugal influences on intrinsic brain stem conduction in cat and monkey. *Journal of Neurophysiology, 20*, 1–16.

Adkins, J., Packwood, J.W., & Marshall, G.L. (1969). Spontaneous alternation and d-amphetamine. *Psychonomic Science, 17*, 167–168.

Aggleton, J.P., & Mishkin, M. (1983a). Memory impairments following restricted medial thalamic lesions in monkeys. *Experimental Brain Research, 52*, 199–209.

Aggleton, J.P., & Mishkin, M. (1983b). Visual recognition impairment following medial thalamic lesions in monkeys. *Neuropsychologia, 21*, 189–197.

Aghajanian, G.K., Foote, W.E., & Sheard, M.H. (1970). Action of psychotogenic drugs on single midbrain raphe neurons. *Journal of Pharmacology and Experimental Therapeutics, 171*, 178–187.

Aghajanian, G.K., & Sheard, M.H. (1968). Behavioral effects of midbrain raphe stimulation—Dependence on serotonin. *Communications in Behavioral Biology, 1*, 37–41.

Aitken, P.P. (1972). Aversive stimulation and rats' preference for familiarity. *Psychonomic Science, 28*, 281–282.

Akre, R.D. (1964). "Correcting" behavior by insects on vertical and horizontal mazes. *Journal of the Kansas Entomological Society, 37*, 169–186.

Albert, D.J., & Walsh, M.L. (1984). Neural systems and the inhibitory modulation of agonistic behavior: A comparison of mammalian species. *Neuroscience and Biobehavior Reviews, 8*, 5–24.

Alberts, J.R. (1981). Ontogeny of olfaction: Reciprocal roles of sensation and behavior in the development of perception. In R.N. Aslin, J.R. Alberts, & M.R. Petersen (Eds.), *Development of perception: Psychobiological perspectives* (Vol. 1, pp. 321–357). New York: Academic Press.

Alberts, J.R. (1984). Sensory-perceptual development in the Norway rat: A view toward comparative studies. In R. Kail & N.E. Spear (Eds.), *Comparative perspectives on the development of memory* (pp. 65–101). Hillsdale, NJ: Erlbaum.

Alberts, J.R., & Friedman, M.I. (1972). Olfactory bulb removal but not anosmia increases emotionality and mouse killing. *Nature, 238*, 454–455.

Aller, T., Hiral, D., & Caldwell, R.L. (1979). A comparison of the correcting behaviour exhibited by two milkweed bugs, *Oncopeltus fasciatus* and *Lygaeus kalmil. Physiological Entomology, 4*, 99–102.

Alpers, B.J. (1937). Relation of the hypothalamus to disorders of personality. *Archives of Neurology and Psychiatry, 38*, 291–303.

Altman, J., & Das, G.D. (1965). Autoradiographic and histological evidence of postnatal hippocampal neurogenesis in rats. *Journal of Comparative Neurology, 124*, 319–336.

Altman, J., & Das, G.D. (1967). Postnatal neurogenesis in the guinea pig. *Nature, 314*, 1098–1101.

Anisman, H. (1975). Dissociation of disinhibitory effects of scopolamine: Strain and task factors. *Pharmacology, Biochemistry and Behavior, 3*, 613–618.

Anisman, H., & Kokkinidis, L. (1975). Effects of scopolamine, d-amphetamine and other drugs affecting catecholimines on spontaneous alternation and locomotor activity in mice. *Psychopharmacologia, 45*, 55–63.

Anisman, H., Kokkinidis, L., Glazier, S., & Remington, G. (1976). Differentiation of response biases elicited by scopolamine and d-amphetamine: Effects on habituation. *Behavioral Biology, 18*, 401–417.

Antelman, S.H., & Caggiula, A.R. (1977). Norepinephrine-dopamine interactions and behavior. *Science, 195*, 646–653.

Arendt, T., Bigl, V., Arendt, A., & Tennstedt, A. (1983). Loss of neurons in the nucleus basalis of Meynert in Alzheimer's disease, paralysis agitans and Korsakoff's disease. *Acta Neuropathologica, 61*, 101–108.

Asin, K.E., & Fibiger, H.C. (1984). Spontaneous and delayed spatial alternation following damage to specific neuronal elements within the nucleus medianus raphe. *Behavioral Brain Research, 13*, 241–250.

Baettig, K. (1963). Effect of lesions on spontaneous alternation and exploration behavior. *Psychological Reports, 13*, 493–494.

Baisden, R.H., Isaacson, R.L., Woodruff, M.L., & Van Hartesveldt, D. (1972). The effect of physostigmine on spontaneous alternation in infant rabbits. *Psychonomic Science, 26*, 287–288.

Baker, J.G., & Franken, R.E. (1968). Spontaneous alternation and the influence of reward. *Psychological Reports, 22*, 821–825.

Ballachey, E.L., & Buel, J. (1934). Centrifugal swing as a determinant of choice-point behavior in the maze running of the white rat. *Journal of Comparative Psychology, 17*, 201–223.

Barlow, J.S. (1964). Inertial navigation as a basis for animal navigation. *Journal of Theoretical Biology, 6*, 76–117.

Barnwell, F.H. (1965). An angle sense in the orientation of a millipede. *Biological Bulletin, 128*, 33–50.

Bate, P., & Kirkby, R.J. (1977). Response alternation in coral reef fish. *Journal of Biological Psychology, 19*, 31–32.

Bayer, S.A., Brunner, R.L., Hine, R., & Altman, J. (1973). Behavioral effects of interference with the postnatal acquisition of hippocampal granule cells. *Nature, New Biology, 242*, 222–224.

Beale, I.L., & Webster, D.M. (1971). The relevance of leg-movement cues to turn alternation in woodlice (*Porcellio scaber*). *Animal Behaviour, 19*, 353–356.

Beracochea, D.J., & Jafford, R. (1987). Impairment of spontaneous alternation behavior in sequential test procedures following mammillary body lesions in mice: Evidence for time-dependent interference-related deficits. *Behavioral Neuroscience, 101*, 187–197.

Beritoff, J.S. (1965). *Neural mechanisms of higher vertebrate behavior* (W. Liberson, Trans., Ed.). Boston: Little Brown & Co.

Beusekom, G. van (1948). Some experiments on the optical orientation in *Philanthrus triangulum Fabr. Behaviour, 1*, 195–226.

Bicker, G., & Spatz, H-C. (1976). Maze-learning ability of *Drosophila melanogaster. Nature, 260*, 371.

Bignami, G. (1976). Nonassociative explanations of behavioral changes induced by central cholinergic drugs. *Acta Neurobiologiae Experimentalis, 36*, 5–90.

Bisiach, E., & Luzzatti, C. (1978). Unilateral neglect of representational space. *Cortex, 14*, 129–133.

Blampied, N.M., & Harvey, S.P. (1979). 3-acetylpyridine and spontaneous alternation: A replication. *IRCS Medical Science, 7*, 261.

Blampied, N.M., & Wilby, C.M. (1975). Effects of 3-acetylpyridine on spontaneous alternation in the mouse. *Pharmacology, Biochemistry and Behavior, 3*, 317–319.

Bliss, T.V.P., & Errington, M.L. (1977). "Reeler" mutant mice fail to show spontaneous alternation. *Brain Research, 124*, 168–170.

Borst, A., Delacour, J., & Libouban, S. (1970). Effets, chez le rat, de lesions du noyau caude sur le conditionment de response alternee. *Neuropsychologia, 8*, 89–101.

Bowen, F.P., Hoehn, M., & Yahr, M.D. (1972). Parkinsonism: Alterations in spatial orientation as determined by a route-walking test. *Neuropsychologia, 10*, 355–361.

Brennan, J.F., & Barone, R.J. (1976). Effects of differential cue availability in an active avoidance conditioned stimulus for young and adult rats. *Developmental Psychobiology, 9*, 237–244.

Brill, M. (1967). *Parameters of odor-trail avoidance in the spontaneous alternation of the rat*. Unpublished doctoral dissertation, University of Cincinnati, OH.

Brion, S. (1969). Korsakoff's syndrome: Clinico-anatomical and physiopathological considerations. In G. Talland & N. Waugh (Eds.), *The pathology of memory* (pp. 29–39). New York: Academic Press.

Brody, J.F. (1970). Behavioral effects of serotonin depletion and of p-chlorophenylalanine (a serotonin depletor) in rats. *Psychopharmacologia, 14*, 14–33.

Bronstein, P.M. (1972a). Open-field behavior of the rat as a function of age: Cross sectional and longitudinal investigations. *Journal of Comparative and Physiological Psychology, 80*, 335–341.

Bronstein, P.M. (1972b). Repeated trials with the albino rat in the open field as a function of age and deprivation. *Journal of Comparative and Physiological Psychology, 81*, 84–93.

Bronstein, P.M., Dworkin, T., & Bilder, B.H. (1974). Age-related difference in rats' spontaneous alternation. *Animal Learning & Behavior, 2*, 285–288.

Bronstein, P.M., Neiman, H., Wolkoff, F.D., & Levine, M.J. (1974). The development of habituation in the rat. *Animal Learning & Behavior, 2*, 92–96.

Brown, C.P. (1976). Two types of habituation in chicks: Differential dependence on cholinergic activity. *Pharmacology, Biochemistry and Behavior, 4*, 235–238.

Bruto, V., Beauchamp, C., Zacharko, R.M., & Anisman, H. (1984). Amphetamine-induced perseverative behavior in a radial arm maze following DSP4 or 6-OHDA pretreatment. *Psychopharmacology, 83*, 62–69.

Burger, M.L. (1971). Zum Mechanismus der Gegenwendung nach mechanisch aufgezwungener Richtungsänderung bei *Schizophyllum sabulosum (Julidae, Diplopoda). Zeitschrift für Vergleichende Physiologie, 71*, 219–254.

Buser, P., Ponderoux, G., & Mereaux, J. (1974). Single unit recording in the caudate nucleus during sessions with elaborate movements in the awake monkey. *Brain Research, 71*, 337–344.

Butcher, L.L., Engel, J., & Fuxe, K. (1972). Behavioral, biochemical and histochemical analyses of the central effects of monoamine precursors after peripheral decarboxylase inhibition. *Brain Research, 41*, 387–395.

Butler, R.A. (1953). Discrimination learning by rhesus monkeys to visual-exploration motivation. *Journal of Comparative and Physiological Psychology, 46*, 95–98.

Caldwell, W.E., & Phaup, M.R. (1959). Some aspects of the behavior of the ant in the maze under various conditions of stimulation. *Journal of Genetic Psychology, 95*, 27–39.

Campbell, B.A., & Coulter, X. (1976). Ontogeny of learning and memory. In M.R. Rosenzweig & E.L. Bennet (Eds.), *Neural mechanisms of learning and memory* (pp. 209–235). Cambridge, MA: MIT Press.

Campbell, B.A., Lytle, L.D., & Fibiger, H.C. (1969). Ontogeny of adrenergic arousal and cholinergic inhibitory mechanisms in the rat. *Science, 166*, 635–637.

Campbell, B.A., & Spear, N.E. (1972). Ontogeny of memory. *Psychological Review, 79*, 215–236.

Campbell, B.A., & Stehower, D.J. (1979). Ontogeny of habituation and sensitization in the rat. In N.E. Spear & B.A. Campbell (Eds). *Ontogeny of learning and memory* (pp. 67–100). Hillsdale, NJ: Erlbaum.

Capaldi, E.J. (1963). Overlearning reversal effect in a spatial discrimination task. *Perceptual and Motor Skills, 16*, 335–336.

Cappa, S., Sterzi, R., Vallar, G., & Bisiach, E. (1987). Remission of hemineglect and anosognosia during vestibular stimulation. *Neuropsychologia, 25*, 775–782.

Carlsson, A. (1970). Amphetamine and brain catecholamines. In E. Costa and S. Garattini (Eds.), *Amphetamines and related compounds* (pp. 289–300). New York: Raven Press.

Carlton, P.L. (1963). Cholinergic mechanisms in the control of behavior by the brain. *Psychological Review, 70*, 19–39.

Carlton, P.L. (1968). Brain acetylcholine and habituation. *Progressive Brain Research, 28*, 48–60.

Carlton, P.L. (1969). Brain-acetylcholine and inhibition. In J.T. Tapp (Ed.), *Reinforcement and behavior* (pp. 286–327). New York: Academic Press.

Carlton, P.L., & Advokat, C. (1973). Attenuated habituation due to parachlorophenylalanine. *Pharmacology, Biochemistry and Behavior, 1*, 657–663.

Carlton, P.L., & Vogel, J.P. (1965). Studies of the amnesic properties of scopolamine. *Psychonomic Science, 3*, 261–262.

Carr, H.A. (1917). The distribution and elimination of errors in the maze. *Journal of Animal Behavior, 7*, 145–159.

Casteigne, P., Lehrmitte, F., Buge, A., Escourolle, R., Hauw, J.J., & Lyon-Caen, O. (1981). Paramedian thalamic and midbrain infarcts: Clinical and neuropathological study. *Annals of Neurology*, *10*, 127–148.

Castellucci, V., & Kandel, E.R. (1976). An invertebrate system for the cellular study of habituation and sensitization. In T.J. Tighe & R.N. Leaton (Eds.), *Habituation: Perspectives from child development, animal behavior, and neurophysiology* (pp. 1–47). Hillsdale, NJ: Erlbaum.

Cerletti, A., & Rothlin, E. (1955). Role of 5-hydroxytryptamine in mental diseases and its antagonism to lysergic acid derivatives. *Nature*, *176*, 155–157.

Clayton, K.N. (1963). Reversal performance by rats following overlearning with and without irrelevant stimuli. *Journal of Experimental Psychology*, *66*, 255–259.

Clody, D.E., & Carlton, P.L. (1969). Behavioral effects of lesions of the medial septum of rats. *Journal of Comparative and Physiological Psychology*, *67*, 344–351.

Cogan, D., Jones, J.F., & Irons, T. (1979). Spontaneous alternation in chicks using social reward. *Developmental Psychobiology*, *12*, 285–290.

Connor, R.L., Stolk, J.M., Barchas, B.D., & Levine, S. (1970). Parachlorophenylalanine and habituation to repetitive auditory startle stimuli in rats. *Physiology and Behavior*, *5*, 1215–1219.

Copack, P., Dafny, N., & Gilman, S. (1972). Neurophysiological evidence of vestibular projections to thalamus, basal ganglia, and cerebral cortex. In T. Frigyesi, E. Rinvik, & M. Yahr (Eds.), *Corticothalamic projections and sensorimotor activities* (pp. 309–335). New York: Raven Press.

Corby, J.R., Caza, P.A., & Spear, N.E. (1982). Ontogenetic changes in the effectiveness of home nest odor as a conditioned stimulus. *Behavioral and Neural Biology*, *35*, 354–367.

Cornwall, J., & Phillipson, O.T. (1988). Afferent projections to the parafascicular thalamic nucleus of the rat, as shown by the retrograde transport of wheat germ agglutin. *Brain Research Bulletin*, *20*, 139–150.

Cotman, C.W., Monaghan, D.R., Ottersen, O.P., & Storm-Mathisen, J. (1987). Anatomical organization of excitatory amino acid receptors and their pathways. *Trends in Neuroscience*, *10*, 273–280.

Cramon, D.Y., von Hebel, N., & Schuri, U. (1985). A contribution to the anatomical basis of thalamic amnesia. *Brain*, *108*, 993–1008.

Critchley, M. (1966). *The parietal lobes*. Baltimore: Williams & Wilkins.

Dalland, T. (1970). Response and stimulus perseveration in rats with septal and dorsal hippocampal lesions. *Journal of Comparative and Physiological Psychology*, *71*, 114–118.

Dalland, T. (1974). Stimulus perseveration of rats with septal lesions. *Physiology & Behavior*, *12*, 1057–1061.

Darchen, R. (1952). Sur l'activité exploratrice de *Blatella germanica*. *Zeitschrift für Tierpsychologie*, *9*, 362–372.

Darchen, R. (1955). Stimuli nouveaux et tendance exploratrice chez *Blatella germanica*. *Zeitschrift für Tierpsychologie*, *12*, 1–11.

Dashiell, J.F., & Bayroff, A.G. (1931). A forward-going tendency in maze-running. *Journal of Comparative Psychology*, *12*, 77–94.

Davis, M. (1974). Sensitization of the startle response by noise. *Journal of Comparative and Physiological Psychology*, *87*, 571–581.

Davis, M. (1980). Neurochemical modulation of sensory-motor reactivity: Acoustic and tactile startle reflexes. *Neuroscience Biobehavioral Reviews*, *4*, 241–263.

Davis, M., & Aghajanian, G.K. (1976). Effects of apomorphine and haloperidol on the acoustic startle response in rats. *Psychopharmacology, 47*, 217–223.

Davis, M., & Astrachan, D.I. (1981). Spinal modulation of acoustic startle: Opposite effects of clonidine and d-amphetamine. *Psychopharmacology, 75*, 219–225.

Davis, M., Astrachan, D.I., & Kass, E. (1980). Excitatory and inhibitory effect of serotonin on sensorimotor reactivity measured with acoustic startle. *Science, 209*, 521–523.

Davis, M., Cedarbaum, J.M., Aghajanian, G.K., & Gendelman, D.S. (1977). Effects of clonidine on habituation and sensitization of acoustic startle in normal, decerebrate and locus coeruleus lesioned rats. *Psychopharmacology, 51*, 243–253.

Davis, M., Commissaris, R.L., Cassella, J.V., Yang, S., Dember, L., & Harty, T.P. (1986). Differential effects of dopamine agonists on acoustically and electrically elicited startle responses: Comparison to effects of strychnine. *Brain Research, 371*, 58–69.

Davis, M., & Sheard, M.H. (1974a). Biphasic dose-response effects of N, N-dimethyltryptamine on the rat startle reflex. *Pharmacology, Biochemistry and Behavior, 2*, 827–836.

Davis, M., & Sheard, M.H. (1974b). Effects of habituation and sensitization of the startle response in the rat. *Pharmacology, Biochemistry and Behavior, 2*, 675–683.

Davis, M., & Sheard, M.H. (1974c). Habituation and sensitization of the rat startle response: Effects of raphe lesions. *Physiology & Behavior, 12*, 425–431.

Davis, M., & Sheard, M.H. (1976). p-Chloroamphetamine (PcA): Acute and chronic effects on habituation and sensitization of the acoustic startle response in rats. *European Journal of Pharmacology, 35*, 261–273.

Davis, M., Svensson, T.H., & Aghajanian, G.K. (1975). Effects of d- and l-amphetamine on habituation and sensitization of the acoustic startle response in rats. *Psychopharmacologia, 43*, 1–11.

Deakin, J.F.W., File, S.E., Hyde, J.R.G., & MacLeod, N.K. (1979). Ascending 5-HT pathways and behavioral habituation. *Pharmacology, Biochemistry and Behavior, 10*, 687–694.

Dember, W.N. (1956). Response by the rat to environmental change. *Journal of Comparative and Physiological Psychology, 49*, 93–95.

Dember, W.N. (1958). Stimulus alternation in peripherally-blinded rats. *Canadian Journal of Psychology, 12*, 219–221.

Dember, W.N. (1960). *Alternation and exploration as indicator responses*. Paper presented at the annual meeting of the American Psychological Association, Chicago, IL.

Dember, W.N. (1961). Alternation behavior. In D.W. Fiske & S.R. Maddi (Eds.), *Functions of varied experience* (pp. 227–252). Homewood, IL: Dorsey Press.

Dember, W.N. (1965). The new look in motivation. *American Scientist, 53*, 409–427.

Dember, W.N., & Dember, J.E. (1983). Spontaneous alternation behavior. In B. Wolman (Ed.), *International encyclopedia of psychiatry, psychology, psychoanalysis, and neurology*. Progress Volume 1. (pp. 433–435). New York: Aesculapius Publishers.

Dember, W.N., & Earl, R.W. (1957). Analysis of exploratory, manipulatory and curiosity behaviors. *Psychological Review, 64*, 91–96.

Dember, W.N., & Fowler, H. (1958). Spontaneous alternation behavior. *Psychological Bulletin, 55*, 412–428.

Dember, W.N., & Kleinman, R. (1973). Cues for spontaneous alternation by gerbils. *Animal Learning & Behavior, 1*, 287–289.

Dember, W.N., & Millbrook, B.A. (1956). Free-choice by the rat of the greater of two brightness changes. *Psychological Reports, 2*, 464–467.

Dember, W.N., & Roberts, W.W. (1958). Alternation behavior in peripherally-blinded rats. *Perceptual and Motor Skills, 8*, 91–94.

Dember, W.N., Sherrick, M.F., & Harris, R.P. Jr. (1966). Trial-two goal arm alternation to orientation of trial-one starting stem. *Psychonomic Science, 6*, 31–32.

Dennis, W. (1935). A comparison of the rat's first and second explorations of a maze unit. *American Journal of Psychology, 47*, 488–490.

Dennis, W. (1939). Spontaneous alternation in rats as an indicator of the persistence of stimulus effects. *Journal of Comparative Psychology, 28*, 305–312.

Dennis, W., & Henneman, R.H. (1932). The non-random character of initial maze behavior. *Journal of Genetic Psychology, 40*, 396–405.

Dennis, W., Henneman, R.H., & Varner, W.B. (1932). A further study of the initial maze behavior of rats. *Journal of Genetic Psychology, 41*, 358–368.

Dennis, W., & Sollenberger, R.T. (1934). Negative adaptation in the maze exploration of albino rats. *Journal of Comparative Psychology, 18*, 197–206.

DeValois, R.C. (1954). The relation of different levels and kinds of motivation to variability of behavior. *Journal of Experimental Psychology, 47*, 392–398.

Diaz, J., Ellison, G., & Masuoka, D. (1974). Opposed behavioral syndromes in rats with partial and more complete central serotonergic lesions made with 5, 6-dihydroxytryptamine. *Psychopharmacologia, 37*, 67–69.

Dingle, H. (1961). Correcting behavior in boxelder bugs. *Ecology, 42*, 207–211.

Dingle, H. (1962). The occurrence of correcting behavior in various insects. *Ecology, 43*, 727–728.

Dingle, H. (1964a). Correcting behaviour in mealworms (*Tenebrio*) and the rejection of a previous hypothesis. *Animal Behaviour, 12*, 137–139.

Dingle, H. (1964b). Further observations on correcting behaviour in boxelder bugs. *Animal Behaviour, 12*, 116–124.

Dingle, H. (1965). Turn alternation by bugs on causeways as a delayed compensatory response and the effects of varying visual inputs and length of straight path. *Animal Behaviour, 13*, 171–177.

Divac, I., Wikmark, R.G.E., & Gade, A. (1975). Spontaneous alternation in rats with lesions of the frontal lobe: An extension of the frontal lobe syndrome. *Physiological Psychology, 3*, 39–42.

Dokla, C.P.J., Kasprow, W.J., Sideleau, M.M., & Boitano, J.J. (1980). Effects of electroconvulsive shock on open-field behavior and spontaneous alternation in rats. *Behavioral and Neural Biology, 28*, 266–284.

Douglas, R.J. (1964). *An analysis of spontaneous alternation cues*. Unpublished doctoral dissertation, University of Michigan, Ann Arbor.

Douglas, R.J. (1966a). Cues for spontaneous alternation. *Journal of Comparative and Physiological Psychology, 62*, 171–183.

Douglas, R.J. (1966b). Spontaneous alternation and middle ear disease. *Psychonomic Science, 4*, 243–244.

Douglas, R.J. (1967). The hippocampus and behavior. *Psychological Bulletin, 67*, 416–442.

Douglas, R.J. (1972). Pavlovian conditioning and the brain. In R. Boakes & M. Halliday (Eds.), *Inhibition and learning* (pp. 529–553). New York: Academic Press.

Douglas, R.J. (1975). The development of hippocampal function: Implications for theory and therapy. In R.L. Isaacson & K.H. Pribram (Eds.), *The Hippocampus. Vol. 2: Neurophysiology and behavior* (pp. 327–361). New York: Plenum Press.

Douglas, R.J., Clark, G.M., Erway, L.C., Hubbard, D.G., & Wright, C.G. (1979). Effects of genetic vestibular defects on behavior related to spatial orientation and emotionality. *Journal of Comparative and Physiological Psychology, 93*, 467–480.

Douglas, R.J., Douglas, D.P., & Treneer, C.M. (1979). *Aging and inhibition* (Final report on Grant 5 R01 AG00013). Washington, DC: National Institute of Health.

Douglas, R.J., & Isaacson, R.L. (1964). Hippocampal lesions and activity. *Psychonomic Science, 1*, 187–188.

Douglas, R.J., & Isaacson, R.L. (1965). Homogeneity of single trial response tendencies and spontaneous alternation in the T-maze. *Psychological Reports, 16*, 87–92.

Douglas, R.J., & Isaacson, R.L. (1966). Spontaneous alternation and scopolamine. *Psychonomic Science, 4*, 283–284.

Douglas, R.J., Isaacson, R.L., & Moss, R.L. (1969). Olfactory lesions, emotionality and activity. *Physiology and behavior, 4*, 379–381.

Douglas, R.J., Kowal, D., & Clark, G. (1968). *Spontaneous alternation and the brain.* Presented at the annual meeting of the Western Psychological Association, San Diego, CA.

Douglas, R.J., McGarvey, E.T., & Clark, G.M. (1984). Alcohol-induced "manic" running in mice. *Society for Neuroscience Abstracts, 10*, 960.

Douglas, R.J., McGarvey, E.T., & Reilly, C. (1987). Inhibition of ethanol-induced locomotion in mice. *Society for Neuroscience Abstracts, 13*, 1723.

Douglas, R.J., Mitchell, D., & delValle, R. (1974). Angle between choice alleys as a critical factor in spontaneous alternation. *Animal Learning & Behavior, 2*, 218–220.

Douglas, R.J., Mitchell, D., & Kentala, D. (1972). Spontaneous alternation as a function of maze configuration. *Psychonomic Science, 27*, 285–286.

Douglas, R.J., Pagano, R.R., Lovely, R.H., & Peterson, J.J. (1973). The prolonged effects of a single ECS on behavior related to hippocampal function. *Behavioral Biology, 8*, 611–617.

Douglas, R.J., Peterson, J.J., & Douglas, D.P. (1973). The ontogeny of a hippocampus-dependent response in two rodent species. *Behavioral Biology, 8*, 27–37.

Douglas, R.J., & Pribram, K.H. (1966). Learning and limbic lesions. *Neuropsychologia, 4*, 197–220.

Douglas, R.J., & Raphelson, A.C. (1966). Spontaneous alternation and septal lesions. *Journal of Comparative and Physiological Psychology, 62*, 320–322.

Douglas, R.J., & Truncer, P. (1976). Parallel but independent effects of pentobarbital and scopolamine on hippocampus-related behavior. *Behavioral Biology, 18*, 359–367.

Drachman, D.A. (1977). Memory and cognitive function in man: Does the cholinergic system have a specific role? *Neurology, 27*, 783–790.

Drew, W.G., Miller, L.C., & Baugh, E.L. (1973). Effects of delta nine-THC, LSD-25 and scopolamine on continuous spontaneous alternation in the Y-maze. *Psychopharmacologia, 32*, 171–182.

Duel, R.K., & Collins, R.C. (1983). Recovery from unilateral neglect. *Experimental Neurology, 81*, 733–748.

Dutch, J. (1969). Visual complexity and stimulus pacing in chicks. *Quarterly Journal of Experimental Psychology, 21*, 63–66.

Eastment, A.M., & Hughes, R.N. (1968). Reactions of ferret-polecat hybrids to complexity and change. *Perceptual and Motor Skills, 26*, 935–938.

Egger, G.J. (1973a). Novelty induced changes in spontaneous alternation by infant and adult rats. *Developmental Psychobiology, 6*, 431–435.

Egger, G.J. (1973b). The relevance of memory, arousal, and cue factors to developmental changes in spontaneous alternation. *Developmental Psychobiology, 6*, 459–468.

Egger, G.J., Livesey, P.J., & Dawson, R.G. (1973). Ontogenetic aspects of central cholinergic involvement in spontaneous alternation behavior. *Developmental Psychobiology, 6*, 289–299.

Einon, D.F., & Morgan, M.J. (1978). Early isolation produces enduring hyperactivity in the rat, but no effect on spontaneous alternation. *Quarterly Journal of Experimental Psychology, 30*, 151–156.

Eisenberger, R., Myers, A.K., Sanders, R., & Shanab, M. (1970). Stimulus control of spontaneous alternation in the rat. *Journal of Comparative and Physiological Psychology, 70*, 136–140.

Ellen, P., & DeLoache, J. (1968). Hippocampal lesions and spontaneous alternation in the rat. *Physiology and Behavior, 3*, 857–860.

Estes, W.K. (1958). Stimulus-response theory of drive. In M.R. Jones (Ed.), *Nebraska symposium on motivation* (pp. 35–69). Lincoln, NE: University of Nebraska Press.

Estes, W.K., & Schoeffler, M.S. (1955). Analysis of variables influencing alternation after forced trials. *Journal of Comparative and Physiological Psychology, 48*, 357–362.

Etienne, A.S. (1980). The orientation of the golden hamster to its nest-site after the elimination of various sensory cues. *Experientia, 36*, 1048–1050.

Etienne, A.S. (1987). The control of short-distance homing in the golden hamster. In P. Ellen & C. Thinus-Blanc (Eds.), *Cognitive processes and spatial orientation in animals and man*. (Vol. 1, pp. 233–251). Boston: Martinus Nijhoff.

Etienne, A.S., Teroni, E., Maurer, R., Portenier, V., & Saucy, F. (1985). Short-distance homing in a small animal: The role of exteroceptive cues and path integration. *Experientia, 41*, 122–125.

Evans, J.A.C., & Thornton, E.W. (1984). Impaired acquisition of DRL operant responding following lesion of the habenular nucleus. *Physiological Psychology, 12*, 220–226.

Fantie, B.D., & Goddard, G.V. (1982). Septal modulation of the population spike in the fascia dentata produced by perforant path stimulation in the rat. *Brain Research, 252*, 227–237.

Fechter, L.P. (1974a). Central serotonin involvement in the elaboration of the startle reaction in rats. *Pharmacology, Biochemistry and Behavior, 2*, 161–171.

Fechter, L.P. (1974b). The effects of L-Dopa, clonidine, and apomorphine on the acoustic startle reaction in rats. *Psychopharmacologia, 39*, 331–334.

Feigley, D.A., & Hamilton, L.W. (1971). Response to novel environment following septal lesions or cholinergic blockade in rats. *Journal of Comparative and Physiological Psychology, 76*, 496–504.

Feigley, D.A., Parsons, P.A., Hamilton, L.W., & Spear, N.E. (1972). Development of habituation to novel environments in the rat. *Journal of Comparative and Physiological Psychology, 79*, 443–452.

Fibiger, H.C. (1979). Drugs and reinforcement mechanisms: A critical review of the catecholamine theory. *Annual Review of Pharmacology & Toxicology, 18*, 37–56.

Fidura, F.G., & Leberer, M.R. (1974). Spontaneous alternation as a function of number of forced-choice responses in the goldfish (*Carassius auratus*). *Bulletin of the Psychonomic Society, 3*, 181–182.

Field, T.D., Rosenstock, J., King, E.C., & Greene, E. (1978). Behavioral role of the mammillary efferent system. *Brain Research Bulletin, 3*, 451–456.

File, S.E. (1985). What can be learned from the effects of benzodiazapines on exploratory behavior? *Neuroscience Biobehavioral Reviews, 9*, 45–54.

File, S.E., & Wardill, A.G. (1975a). The reliability of the hole board apparatus. *Psychopharmacologia*, *44*, 47–51.

File, S.E., & Wardill, A.G. (1975b). The validity of the head-dipping response as a measure of exploration in a modified hole board. *Psychopharmacologia*, *44*, 53–59.

Fink, J.P., & Smith, G.R. (1980). Mesolimbicocortical dopamine terminal fields are necessary for normal locomotor and investigatory exploration in rats. *Brain Research*, *199*, 359–384.

Fog, R. (1969). Stereotyped and non-stereotyped behavior in rats induced by various stimulant drugs. *Psychopharmacologia*, *14*, 299–304.

Foote, S.L., & Bloom, F.E. (1978). Activity of norepinephrine-containing locus coeruleus neurons in the unanesthetized squirrel monkey. In E. Usdin, I.J. Kopin, & J. Barchas (Eds.), *Catecholamine: Basic and clinical frontiers* (Vol. I, pp. 625–630). New York: Pergamon Press.

Fowler, H. (1958). Response to environmental change: A positive replication. *Psychological Reports*, *4*, 506.

Fowler, H., Blond, J., & Dember, W.N. (1959). Alternation behavior and learning: The influence of reinforcement magnitude, number and contingency. *Journal of Comparative and Physiological Psychology*, *52*, 609–614.

Fowler, H., Fowler, D.E., & Dember, W.N. (1959). The influence of reward on alternation behavior. *Journal of Comparative and Physiological Psychology*, *56*, 220–224.

Franken, R.E., & Baker, J.G. (1969). The effects of drive level on cues utilized in spontaneous alternation. *Psychonomic Science*, *16*, 239–240.

Frederickson, C.J., & Frederickson, M.H. (1979). Emergence of spontaneous alternation in the kitten. *Developmental Psychobiology*, *12*, 615–621.

Frederickson, J.M., Figge, U., Scheid, P., & Kornhuber, H. H. (1966). Vestibular nerve projection to the cerebral cortex of the rhesus monkey. *Experimental Brain Research*, *2*, 318–327.

Freedman, M., & Oscar-Berman, M. (1986). Bilateral frontal lobe disease and selective delayed response deficits in humans. *Behavioral Neuroscience*, *100*, 337–342.

Freeser, H.R., & Raskin, L.A. (1982). Effects of neonatal dopamine depletion on spatial ability during ontogeny. *Behavioral Neuroscience*, *101*, 812–818.

Friberg, L., Olsen, T.S., Roland, P.E., Paulson, O.B., & Lassen, N.A. (1985). Focal increase in blood flow in the cerebral cortex of man during vestibular stimulation. *Brain*, *108*, 609–623.

Fuller, R.W., Hines, C.W., & Mills, J. (1965). Lowering of brain serotonin levels by p-chloramphetamine. *Biochemistry and Pharmacology*, *14*, 483–488.

Galey, D., Durkin, T., Sifakis, G., Kempf, E., & Jaffard, R. (1985). Facilitation of spontaneous and learned spatial behaviors following 6-hydroxydopamine lesions of the lateral septum: A cholinergic hypothesis. *Brain Research*, *340*, 171–174.

Galey, D., Jaffard, R., & LeMoal, M. (1976). Spontaneous alternation disturbance after lesions of the ventral mesencephalic tegmentum in the rat. *Neuroscience Letters*, *3*, 65–69.

Galey, D., Jaffard, R., & LeMoal, M. (1979). Alternation behavior, spatial discrimination, and reversal after electrocoagulation of the ventral mesencephalic tegmentum in the rat. *Behavioral and Neural Biology*, *26*, 81–88.

Gallager, D.W., & Aghajanian, G.K. (1975). Effects of chlorpromazine and lysergic acid diethylamide on efflux of precursor-formed $^{3}$H-serotonin: Correlations with serotonergic impulse flow. *Journal of Pharmacology and Experimental Therapeutics*, *193*, 785–792.

Gerjuoy, I.R., & Winters, J.J. (1968). Development of lateral and choice-sequence preferences. In N.R. Ellis (Ed.), *International review of research in mental retardation* (Vol. 3, pp. 31-63). New York: Academic Press.

Geyer, M.A. (1975). Functional heterogeneity within neurotransmitter systems. *Psychopharmacology Communications*, *1*, 675-686.

Geyer, M.A. (1978). Heterogeneous functions of discrete serotonergic pathways in brain. In E. Usdin & A.J. Mandell (Eds.), *New vistas in the biochemistry of mental disorders* (Vol. 13, pp. 233-257). New York: Marcel Dekker.

Geyer, M.A., Puerto, A., Menkes, D.B., Segal, D.S., & Mandell, A.J. (1976). Behavioral studies following lesions of the mesolimbic and mesostriatal serotonergic pathways. *Brain Research*, *106*, 257-270.

Gibbs, R.B., Yu, J., & Cotman, C.W. (1987). Entorhinal transplants and spatial memory abilities in rats. *Behavioral Brain Research*, *26*, 29-35.

Glanzer, M. (1953a). Stimulus satiation: An explanation of spontaneous alternation and related phenomena. *Psychological Review*, *60*, 257-268.

Glanzer, M. (1953b). The role of stimulus satiation in spontaneous alternation. *Journal of Experimental Psychology*, *45*, 387-393.

Gloor, P., Olivier, A., Quesney, L.F., Andermann, F., & Horowitz, S. (1982). The role of the limbic system in experiential phenomena of temporal lobe epilepsy. *Annals of Neurology*, *12*, 129-144.

Gottlieb, G. (1984). Evolutionary trends and evolutionary origins: Relevance to theory in comparative psychology. *Psychological Review*, *91*, 448-456.

Graff-Radford, N.R., Tranel, D., & Brandt, J. (1988). Anatomical and neuropsychological characteristics of amnesia from bilateral thalamic infarcts. *Neurology*, *38*, 171.

Grandjean, E., & Battig, K. (1962). Die Wirkung verscheidener Psychopoharmaka auf die spontane Alternation der Ratte. *Helvetica Physioslogia Pharmacologia Acta*, *20*, 373-381.

Grastyan, E. (1959). The hippocampus and higher nervous activity. In M. Brazier (Ed.), *The central nervous system and behavior* (pp. 119-205). New York: Josiah Macy, Jr., Foundation.

Gray, J.A., & MacNaughton, N. (1983). Comparison between the behavioral effects of septal and hippocampal lesions: A review. *Neuroscience and Biobehavior Reviews*, *7*, 119-188.

Graybiel, A.M. (1984). Neurochemically specified subsystems in the basal ganglia. In *Ciba Foundation symposium 107* (pp. 114-144). London: Pitman.

Green, S.E., Joyce, D., & Summerfield, A. 91975). Effects of scopolamine on habituation of exploratory activity in rats. *Physiological Psychology*, *3*, 940, 400-404.

Greenberg, G. (1973). Replication report: No spontaneous alternation in gerbils. *Bulletin of the Psychonomic Society*, *1*, 141-142.

Greenberg, G. (1978). Spontaneous alternation in the Mongolian gerbil: A methodological comment. *Journal of General Psychology*, *98*, 271-279.

Greene, E., & Naranjo, J.N. (1986). Thalamic role in spatial memory. *Behavioral Brain Research*, *19*, 123-131.

Griffiths, D., & Whalsten, D. (1974). Interacting effects of handling and d-amphetamine on avoidance learning. *Pharmacology, Biochemistry and Behavior*, *2*, 439-441.

Groenewegen, H.J., Vermeulen-Van der Zee, E., Kortschot, A.T.E., & Witter, M.P. (1987). Organization of the projections from the subiculum to the ventral striatum in the rat. A study using anterograde transport of *Phaseolus vulgaris* leucoagglutinin. *Neuroscience*, *23*, 103-120.

Grosslight, J.H., & Harrison, P.C. (1961). Variability of response in a determined turning sequence in the meal worm (*Tenebrio molitor*): An experimental test of alternative hypotheses. *Animal Behaviour, 9*, 100–103.

Grosslight, J.H., & Ticknor, W. (1953). Variability and reactive inhibition in the meal worm as a function of determined turning sequences. *Journal of Comparative and Physiological Psychology, 46*, 35–38.

Groves, P.M., & Lynch, G.S. (1972). Mechanisms of habituation in the brain stem. *Psychological Review, 79*, 237–244.

Groves, P.M., & Thompson, R. (1970). Habituation: A dual process theory. *Psychological Review, 77*, 419–450.

Groves, P.M., & Thompson, R.F. (1973). A dual-process theory of habituation neural mechanisms. In H.V.S. Peeke and M.J. Herz (Eds.), *Habituation* (Vol. II, pp. 203–207). New York: Academic Press.

Groves, P.M., Wilson, C.J., & Boyle, R.D. (1974). Brain stem pathways, cortical modulation, and habituation of the acoustic startle response. *Behavioral biology, 10*, 391–418.

Halgren, E., Walter, R.D., Cherlow, D.G., & Crandall, P.H. (1978). Mental phenomena evoked by electrical stimulation of the human hippocampal formation and amygdala. *Brain, 101*, 83–117.

Hammond, E., Meador, K.J., Aung-Din, R., & Wilder, B.J. (1987). Cholinergic modulation of human P3 event-related potentials. *Neurology, 37*, 346–350.

Handley, S.L., & Thomas, K.V., (1979). Potentiation of startle response by d- and l-amphetamine: The possible involvement of pre-and postsynaptic alpha-adrenoceptors and other transmitter systems in the modulation of a tactile startle response. *Psychopharmacology, 64*, 105–111.

Harlow, H.F. (1950). Learning and satiation of response in intrinsically motivated complex puzzle performance by monkeys. *Journal of Comparative and Physiological Psychology, 43*, 289–294.

Hartman, A.M. (1982). Acquisition of win-shift predation feeding in rainbow trout: Influence of spatial alternation behaviors. *Journal of Comparative and Physiological Psychology, 96*, 1007–1011.

Hayes, W.N., & Warren, J.M. (1963). Failure to find spontaneous alternation in chicks. *Journal of Comparative and Physiological Psychology, 56*, 575–577.

Haywood, H.G., & Wachs, T.D. (1967). Effects of arousing stimulation upon novelty preference in rats. *British Journal of Psychology, 58*, 77–84.

Healton, E.B., Navarro, C., Bressman, S., & Brust, J.C.M. (1982). Subcortical neglect. *Neurology, 32*, 776–778.

Heathers, G.L. (1940). The avoidance of repetition of a maze reaction in the rat as a function of the time interval between trials. *Journal of Psychology, 10*, 359–380.

Heilizer, F. (1978). Approach-withdrawal response competition displacement, and behavior modification. *Journal of General Psychology, 99*, 181–204.

Heise, G.A., Hrabrich, B., Lilie, N.L., & Martin, R.A. (1975). Scopolamine effects in delayed spatial alternation in the rat. *Pharmacology, Biochemistry and Behavior, 3*, 993–1002.

Heller, A., & Moore, R.Y. (1968). Effect of central nervous system lesion on brain monoamines in the rat. *Journal of Pharmacology and Experimental Therapeutics, 150*, 1–9.

Henderson, N.D. (1970). A genetic analysis of spontaneous alternation in mice. *Behavior Genetics, 1*, 125–132.

Herkenham, M. (1978). The connections of the nucleus reuniens thalami: Evidence for a direct thalamo-hippocampal pathway in the rat. *Journal of Comparative Neurology, 177*, 589–610.

Herkenham, M,. & Nauta, W.J.H. (1979). Efferent connections of the habenular nuclei in the rat. *Journal of Comparative Neurology, 187*, 19–49.

Hess, C., & Blozovski, D. (1987). Hippocampal muscarinic cholinergic mediation of spontaneous alternation and fear in the developing rat. *Behavioral Brain Research, 24*, 203–214.

Hicks, L.H. (1964). Effects of overtraining on acquisition and reversal of place and response learning. *Psychological Reports, 15*, 459–462.

Hill, A.J., & Best, P.J. (1981). Effects of deafness and blindness on the spatial correlates of hippocampal unit activity in the rat. *Experimental Neurology, 74*, 204–217.

Hinde, R.A. (1954a). Factors governing the changes in strength of a partially inborn response, as shown by mobbing behavior of the chaffinch (*Fringilla coelebs*). I. The nature of the response, and an examination of its course. *Proceedings of the Royal Society Serial, B142*, 306–331.

Hinde, R.A. (1954b). Factors governing the changes in strength of a partially inborn response, as shown by mobbing behavior of the chaffinch (*Fringilla coelebs*), II. The waning of the response. *Proceedings of the Royal Society Serial, B142*, 306–331.

Hinde, R.A. (1961). Factors governing the changes in strength of a partially inborn response, as shown by mobbing behavior of the chaffinch (*Fringilla coelebs*), III. The interaction of short term and long term incremental and decremental effects. *Proceedings of the Royal Society Serial, B153*, 398–420.

Hofer, M.A. (1981). *The roots of human behavior.* San Francisco: W.H. Freeman.

Holson, R. (1986). *An evaluation of the lesion technique as exemplified by the spontaneous alternation deficit resulting from hippocampal lesions.* Unpublished manuscript.

Horn, K.M., & Miller, S.W. (1982). Visual-vestibular interactions in single units of rat vestibular nuclei. *Society for Neuroscience Abstracts, 8*, 939.

Hosoda, K. (1963). Spontaneous alternation in monkeys. *Annual of Animal Psychology, 13*, 61–70.

Hovestadt, A., deJong, G.J., & Meerwaldt, J.D. (1987). Spatial disorientation as an early symptom of Parkinson's disease. *Neurology, 37*, 485–487.

Hughes, R.N. (1964). Responses by the ferret to brightness change. *British Journal of Psychology, 55*, 463–468.

Hughes, R.N. (1965). Spontaneous alternation and response to stimulus change in the ferret. *Journal of Comparative and Physiological Psychology, 60*, 149–150.

Hughes, R.N. (1966a). Some observations of correcting behaviour in woodlice (*Porcellio scaber*). *Animal Behaviour, 14*, 319.

Hughes, R.N. (1966b). Stimulus-change effects of forcing in alternation behavior. *Psychological Reports, 19*, 515–518.

Hughes, R.N. (1967a). Effects of intertrial confinement and brightness change on ferret alternation behavior. *Psychological Reports, 20*, 679–681.

Hughes, R.N. (1967b). Turn alternation in woodlice (*Porcellio scaber*). *Animal Behaviour, 15*, 282–286.

Hughes, R.N. (1973). Spontaneous alternation in rabbits. *Bulletin of the Psychonomic Society, 2*, 2.

Hughes, R.N. (1978). Effects of blinding, antennectomy, food deprivation and simulated natural conditions on alternation in woodlice (*Porcellio scaber*). *Journal of Biological Psychology, 20*, 35–40.

Hughes, R.N. (1982). A review of atropinic effects on exploratory choice behavior in laboratory rodents. *Behavioral Neural Biology, 34*, 5–41.

Hughes, R.N. (1984). Cholinergic and monoaminergic substrates of startle habituation. *Life Sciences, 34*, 2101–2105.

Hughes, R.N. (1985). Mechanisms for turn alternation in woodlice (*Porcellio scaber*): The role of bilaterally asymmetrical leg movements. *Animal Learning & Behavior, 13*, 253–260.

Hughes, R.N. (1987). Mechanisms for turn alternation in four invertebrate species. *Behavioural Processes, 14*, 89–103.

Hughes, R.N. (1988). Spontaneous alternation behavior in hamsters. *Current Psychological Research and Reviews, 6*, 331–336.

Hughes, R.N., Blampied, N.M., & Stewart, W.J. (1975). Scopolamine induced changes in activity and reactions to novelty. *Pharmacology, Biochemistry and Behavior, 3*, 731–734.

Hughes, R.N., & Daley, V.A. (1977). Cholinergic involvement in spontaneous alternation. *IRCS Medical Science, 5*, 94.

Hughes, R.N., & Greig, A.M. (1975). Spontaneous alternation in ferrets following treatment with scopolamine, chlordiazepoxide, and caffeine. *Physiological Psychology, 3*, 155–156.

Hughes, R.N., & Seanberg, K.M. (1970). Forcing, stimulus-change, and alternation. *Perceptual and Motor Skills, 31*, 902.

Hull, C.L. (1943). *Principles of behavior.* New York: Appleton-Century-Crofts.

Hunter, M.A., & Ames, E.W. (1988). A multifactor model of infant preferences for novel and familiar stimuli. In C. Rovee-Collier and L.P. Lipsitt (Eds.), *Advances in infancy research* (Vol. 5, pp. 69–95). Norwood, NJ: Ablex Publishing Company.

Hunter, W.S. (1914). The auditory sensitivity of the white rat. *Journal of Animal Behavior, 4*, 215–222.

Irwin, F.W., & Preston, M.G. (1937). Avoidance of repetition of judgments across sense modalities. *Journal of Experimental Psychology, 21*, 511–520.

Isaacson, R.L. (1974). *The limbic system*. New York: Plenum Press.

Isaacson, R.L. (1984). Hippocampal damage: Effects on dopaminergic systems of the basal ganglia. *International Review of Neurobiology, 25*, 339–359.

Isseroff, A. (1979). Limited recovery of spontaneous alternation after extensive hippocampal damage: Evidence for a memory impairment. *Experimental Neurology, 64*, 284–294.

Isseroff, A., & Isseroff, R.G. (1978). Experience aids recovery of spontaneous alternation following hippocampal damage. *Physiology and Behavior, 21*, 469–472.

Isseroff, A., Rosvold, H.E., Galkin, T.W., & Goldman-Rakic, P.S. (1982). Spatial memory impairments following damage to the mediodorsal nucleus of the thalamus in rhesus monkeys. *Brain Research, 232*, 97–113.

Iwahara, S. (1963). Inhibition vs. thigmotropism vs. centrifugal swing as determinants of the initial turn alternation phenomenon in *Armadillidium vulgare. Annual of Animal Psychology, 13*, 1–15.

Iwahara, S., & Fujita, O. (1965). Effect of intertrial interval and removal of the supra-pharyngeal ganglion upon spontaneous alternation in the earthworm, *Pheretima communissima. Japanese Psychological Research, 7*, 1–14.

Iwahara, S., Oishi, H., Yamazaki. S., & Sakai, K. (1972). Effects of chlordiazepoxide upon spontaneous alternation and hippocampal electrical activity in white rats. *Psychopharmacology, 25*, 496–507.

Iwahara, S., & Soeda, N. (1957). The effect of electric shock on spontaneous alternation in the cockroach. *Annual of Animal Psychology, 7*, 43–51.

Iwai, E., & Yukie, M. (1988). A direct projection from hippocampal field CAI to ventral area TE of inferotemporal cortex in the monkey. *Brain Research, 444*, 397–401.

Iwata, K., & Watanabe, M. (1957a). Alternate turning response of *Armadillidium vulgare*. II. Turning and straight going tendencies. *Annual of Animal Psychology, 7*, 535–556.

Iwata, K., & Watanabe, M. (1957b). Alternate turning response of *Armadillidium vulgare*. III. Effect of preceding turning. *Annual of Animal Psychology, 17*, 57–60.

Izquierdo, I., & Cavalheiro, E.A. (1976). The influence of stimulus pairing and of the shuttle-shock contingency in the performance of shuttle response to a buzzer by weanling rats. *Behavioral Biology, 17*, 119–122.

Jackson, M.M. (1941). Reactive tendencies of the white rat in running and jumping situations. *Journal of Comparative Psychology, 31*, 255–262.

Jacobsen, C.F. (1935). Function of the frontal association area in primates. *Archives of Neurology and Psychiatry, 33*, 558–569.

Jacobson, A.L. (1963). Learning in flatworms and annelids. *Psychological Bulletin, 60*, 74–94.

Jensen, D. (1959). A theory of the behavior of *Paramecium aurelia* and behavioral effects of feeding, fission, and ultraviolet microbeam irradiation. *Behaviour, 15*, 82–122.

Jerussi, T.P., & Glick, S.D. (1974). Amphetamine-induced rotation in rats without lesions. *Neuropharmacology, 13*, 238–286.

Johns, J.M., Louis, T.M., Becker, R.F., & Means, L.W. (1982). Behavioral effects of prenatal exposure to nicotine in guinea pigs. *Neurobehavioral Toxicology and Teratology, 4*, 365–369.

Johnson, C.T., Olton, D.S., Gage, F.H. III, & Jenko, P.G. (1977). Damage to hippocampus and hippocampal connections: Effects on DRL and spontaneous alternation. *Journal of Comparative and Physiological Psychology, 91*, 508–522.

Johnson, F.N. (1980). The effects of lithium chloride on spontaneous alternation behaviour in the goldfish (*Carassius auratus*). *Neuropsychobiology, 6*, 72–78.

Jones, G., Segal, M., & Bloom, F. (1978). Locus coeruleus neurons in freely moving rats exhibit pronounced alterations of discharge rate during sensory stimulation and stages of the sleep-wake cycle. In E. Usdin, I.H. Kopin, & J. Barchas (Eds.), *Catecholamines: Basic and clinical frontiers* (Vol. I, pp. 643–645). New York: Pergamon Press.

Jones, S.J. (1970). Children's two-choice learning of predominantly alternating and predominantly non-alternating sequences. *Journal of Experimental Child Psychology, 10*, 344–362.

Joyce, J.N., Davis, R.E., & Hartesfeldt, C.V. (1981). Behavioral effects of unilateral dopamine injection into dorsal or ventral striatum. *European Journal of Pharmacology, 72*, 1–10.

Kaada, B.R. (1951). Somato-motor, autonomic and electrocorticographic responses to electrical stimulation of "rhinencephalic" and other structures in primate, cat, and dog. *Acta Physiologica Scandinavica, 24*, Suppl. 83, 1–258.

Kacelnik, A. (1979). *Studies of foraging behavior and time budgeting in great tits (parus major)*. Unpublished doctoral dissertation, Oxford University, England.

Kail, R., & Spear, N.E. (1984). *Comparative perspectives on the development of memory*. Hillsdale, NJ: Erlbaum.

Kamil, A.C., & Sargent, T.D. (1981). *Foraging behavior: Ecological, ethological and psychological approaches*. New York: Garland Press.

Karpiak, S.E., Rapport, M., & Bowen, F.P. (1974). Immunologically induced behavioral and electrophysiological changes in the rat. *Neuropsychologia, 12*, 313–322.

Katz, R.J., & Schmaltz, K. (1981). Dopaminergic involvement in attention: A novel animal model. *Progress in Neuropsychopharmacology, 4*, 485–490.

Kehne, J.H., & Sorenson, C.A. (1978). The effects of pimozide and phenozybenzamine pretreatments on amphetamine and apomorphine potentiation of the acoustic startle response in rats. *Psychopharmacology, 58*, 137–142.

Kelley, A.E., Winnock, M., & Stinus, L. (1986). Amphetamine, apomorphine and investigatory behavior in the rat: Analysis of the structure and pattern of responses. *Psychopharmacology, 88*, 66–74.

Ketchum, J.S. O, Sidell, F.R., Crowell, E.B. Jr., Aghajanian, G.K., & Hayes, A.H. Jr. (1973). Atropine, scopolamine and ditran: Comparative pharmacology and antagonists in man. *Psychopharmacologia, 28*, 121–145.

Keyes, J.B. (1973). The effects of ECS upon the hippocampus. *Physiological Psychology, 1*, 357–360.

Kim, C., Chang, H.K., & Chu, J.W. (1967). Consequences of ablating the mammillary bodies in dogs. *Journal of Comparative and Physiological Psychology, 63*, 469–476.

Kimble, D.P. (1968). Hippocampus and internal inhibition. *Psychological Bulletin, 70*, 285–295.

Kimble, D.P. (1975). Choice behavior in rats with hippocampal lesions. In R. Isaacson & K. Pribram (Eds.), *The hippocampus* (Vol. 2, pp. 309–326). New York: Plenum Press.

Kimble, D.P. (1976). Changes in behavior of hippocampal-lesioned rats across a 6-week postoperative period. *Physiological Psychology, 4*, 289–293.

Kinghorn, E.W., Fleming, D.E., & Anderson, R.H. (1987). The effects of bilateral and unilateral frontal lesions on visual cortical hypersynchronous bursting and behavioral activity. *Physiology and Behavior, 39*, 297–301.

Kirkby, R.J. (1967). A maturation factor in spontaneous alternation. *Nature, 215*, 784.

Kirkby, R.J. (1969). Caudate nucleus lesions impair spontaneous alternation. *Perceptual and Motor Skills, 29*, 550.

Kirkby, R.J., & Lackey, G.H. (1968). Spontaneous alternation in *Mescricetus auratus*: Age differences. *Psychonomic Science, 10*, 257–258.

Kirkby, R.J., & Preston, A.C. (1972). The behavior of marsupials, II. Reactivity and habituation to novelty in *Sminthopsis crassicaudata. Journal of Biological Psychology, 14*, 21–24.

Kirkby, R.J., Stein, D.G., Kimble, R.J., & Kimble, D.P. (1967). Effects of hippocampal lesions and duration of sensory input on spontaneous alternation. *Journal of Comparative and Physiological Psychology, 64*, 342–345.

Kivy, P.N., Earl, R.W., & Walker, E.L. (1956). Stimulus context and satiation. *Journal of Comparative and Physiological Psychology, 49*, 90–92.

Kleerekoper, H., Timms, A.M., Westlake, G.F., Davy, F.B., Malar, T., & Anderson, V.M. (1970). An analysis of locomotor behaviour of goldfish (*Carassius auratus*). *Animal Behaviour, 18*, 317–330.

Kohler, C. (1976). Habituation of the orienting response after medial and septal lesions in the albino rat. *Behavior Biology, 16*, 63–72.

Kohler, C., & Steinbusch, H. (1982). Identification of serotonin and non-serotonin-containing neurons of the midbrain raphe projection to the entorhinal area and the hippocampal formation. A combined immunohistochemical and fluorescent retrograde study in the rat brain. *Neuroscience, 7*, 951–975.

Kokkinidis, L. (1986). Sensitization to amphetamine and tolerance to cocaine and phencyclidine stimulation in mice. *Pharmacology, Biochemistry and Behavior, 25*, 1175–1180.

Kokkinidis, L. (1987). Amphetamine-elicited perseverative and rotational behavior: Evaluation of directional preference. *Pharmacology, Biochemistry and Behavior, 26*, 527-532.

Kokkinidis, L., & Anisman, H. (1976a). Dissociation of the effects of scopolamine and d-amphetamine in a spontaneous alternation task. *Pharmacology, Biochemistry and Behavior, 5*, 293-297.

Kokkinidis, L., & Anisman, H. (1976b). Interaction between cholinergic and catecholaminergic agents in a spontaneous alternation task. *Psychopharmacology, 48*, 261-270.

Kokkinidis, L., & Anisman, H. (1977). d-Amphetamine-induced perseveration in a Y-maze exploratory task: Differential effects of intraperitoneal and unilateral intraventricular administration. *Psychopharmacology, 52*, 123-128.

Kokkinidis, L., & Anisman, H. (1978a). Abatement of stimulus perseveration following repeated d-amphetamine treatment: Absence of behaviorally augmented tolerance. *Pharmacology, Biochemistry and Behavior, 8*, 557-563.

Kokkinidis, L., & Anisman, H. (1978b). Involvement of norepinephrine in startle arousal after acute and chronic d-amphetamine administration. *Psychopharmacology, 59*, 285-292.

Kokkinidis, L., & Anisman, H. (1980). Amphetamine models of paranoid schizophrenia: An overview and elaboration of animal experimentation. *Psychological Bulletin, 88*, 551-579.

Kokkinidis, L., & Anisman, H. (1981). Amphetamine psychosis and schizophrenia: A dual model. *Neuroscience and Biobehavioral Reviews, 5*, 449-461.

Kokkinidis, L., Irwin, J., & Anisman, H. (1979). Ontogenetic variations in amphetamine-induced stimulus perseveration. *Behavioral and Neural Biology, 26*, 221-233.

Kokkinidis, L., & MacNeill, E.P. (1982a). Potentiation of d-amphetamine and L-Dopa-induced acoustic startle activity after long-term exposure to amphetamine. *Psychopharmacology, 78*, 331-335.

Kokkinidis, L., & MacNeill, E.P. (1982b). Stress induced facilitation of acoustic startle after d-amphetamine administration. *Pharmacology, Biochemistry and Behavior, 17*, 413-417.

Koppenaal, R.J. (1962). On the determination of expected probabilities of alternation. *Psychological Reports, 11*, 666.

Kotchabhakdi, N., & Walberg, F. (1978). Primary vestibular afferent projections to the cerebellum as demonstrated by retrograde axonal transport of horseradish peroxidase. *Brain Research, 142*, 142-146.

Krieckhaus, E.E. (1964). Decrements in avoidance behavior following mammillothalamic tractotomy in cats. *Journal of Neurophysiology, 27*, 753-767.

Kreickhaus, E.E., Coons, E.E., Greenspoon, T., Weiss, J. & Lorenz, R.L. (1968). Retention of choice behavior in rats following mammillothalamaic tractotomy. *Physiology and Behavior, 3*, 125-131.

Kritchevsky, M., Graff-Radford, N.R., & Damasio, A.R. (1987). Normal memory after damage to medial thalamus. *Archives of Neurology, 44*, 959-962.

Kupfermann, I. (1966). Turn alternation in the pill bug (*Armadillidium vulgare*). *Animal Behaviour, 14*, 68-72.

Labuszewski, T., Lockwood, R., McManus, F.E., Edelstein, L.R., & Lidsky, T.I. (1981). Role of postural deficits in oro-ingestive problems caused by globus pallidus lesions. *Experimental Neurology, 74*, 93-110.

Lachman, S.J., & Havlena, J.M. (1962). Reactive inhibition in the paramecium. *Journal of Comparative and Physiological Psychology, 55*, 972-973.

Lahue, R., Kokkinidis, L., & Corning, W. (1975). Telson reflex habituation in *Limulus Polyphemus. Journal of Comparative and Physiological Psychology, 88*, 1061-1069.

Lalonde, R. (1986). Delayed spontaneous alternation in weaver mutant mice. *Brain Research, 398*, 178-180.

Lalonde, R., Botez, M.I., & Boivin, D. (1986). Spontaneous alternation and habituation in a t-maze in nervous mutant mice. *Behavioral Neuroscience, 100*, 350-352.

Lalonde, R., Lamarre, Y., Smith, A.M., & Botez, M.I. (1986). Spontaneous alternation and habituation in lurcher mutant mice. *Brain Research, 362*, 161-164.

Lalonde, R., Manseau, M., & Botez, M.I. (1987). Delayed spontaneous alternation in Purkinje cell degeneration mutant mice. *Neuroscience Letters, 80*, 343-346.

Lalonde, R., Manseau, M., & Botez, M.I. (1988). Spontaneous alternation and exploration in staggerer mutant mice. *Behavioral Brain Research, 27*, 273-276.

Lash, L. (1964). Response discriminability and the hippocampus. *Journal of Comparative and Physiological Psychology, 57*, 251-256.

Laughlin, N.K., Finger, S., & Bell, J. (1984). Early undernutrition and later hippocampal damage: Effects on spontaneous behaviors and reversal learning. *Physiological Psychology, 11*, 269-277.

Leaton, R.N., & Cassella, J.V. (1984). The effects of clonidine, prazosin, and propranolol on short-term and long-term habituation of the acoustic startle response in rats. *Pharmacology, Biochemistry and Behavior, 20*, 935-942.

Leaton, R.N., & Utell, M.J. (1970). Effects of scopolamine on spontaneous alternation following free and forced trials. *Physiology and Behavior, 5*, 331-334.

LeMoal, M., Cardo, B., & Stinus, L. (1969). Influence of ventral mesencephalic lesions on various spontaneous and conditioned behaviors in the rat. *Physiology and Behavior, 4*, 567-573.

LeMoal, M., Stinus, L., & Galey, D. (1976). Radiofrequency lesions of the ventral mesencephalic tegmentum: Neurological and behavioral considerations. *Experimental Neurology, 50*, 521-535.

Leon, M. (1974). Maternal pheromone. *Physiology and Behavior, 13*, 441-453.

Lepley, W.M., & Rice, G.E. (1952). Behavior variability in paramecia as a function of guided act sequences. *Journal of Comparative and Physiological Psychology, 45*, 283-286.

Lester, D. (1968). Response alternation: A review. *Journal of Psychology, 69*, 131-142.

Liljequist, R., Linnoila, M., & Mattila, M.J. (1978). Effect of diazepam and chlorpromazine on memory functions in man. *European Journal of Clinical Pharmacology, 13*, 339-343.

Lindauer, M. (1952). Ein Beitrag zur Frage der Arbeitsteilung im Bienenstaat. *Zeitschrift für Vergleichende Physiologie, 34*, 299-345.

Loesche, J., & Steward, O. (1977). Behavioral correlates of denervation and reinnervation of the hippocampal formation of the rat: Recovery of alternation performance following unilateral entorhinal cortex lesions. *Brain Research Bulletin, 2*, 31-39.

Lovejoy, E. (1966). Analysis of the overlearning reversal effect. *Psychological Review, 73*, 87-103.

Lubow, R.E. (1973). Latent inhibition, *Psychological Bulletin, 79*, 398-407.

Luh, C.W. (1931). The avoidance of repetitive responses. *Psychological Review, 38*, 276-279.

Mabry, P.D., & Campbell, B.A. (1974). Ontogeny of serotonergic inhibition of behavior arousal in the rat. *Journal of Comparative and Physiological Psychology, 86*, 193–201.

MacGillivray, M.E., & Stone, C.P. (1930). Suggestions toward and explanation of systematic errors made by albino rats in multiple light discrimination apparatus. *Journal of Genetic Psychology, 38*, 484–489.

MacKay, D.C.G. (1945). Left-right tendency in the hermit crab, *Calcinus herbsteii. Journal of Comparative Psychology, 38*, 131–133.

MacKay, D.C.G. (1947). Left-right tendency in the hermit crab, *Clibinarius zebra dana. Journal of Comparative and Physiological Psychology, 40*, 421–425.

Mackintosh, N.J. (1965). Overtraining, transfer to proprioceptive control and position reversal. *Quarterly Journal of Experimental Psychology, 17*, 26–36.

MacNaughton, N., & Mason, S.T. (1980). The neuropsychology and neuropharmacology of the dorsal ascending noradrenergic bundle: A review. *Progress in Neurobiology, 14*, 157–219.

Manning, S.K. (1973a). A design comparing free- and forced-trial alternation in the rat at two interchoice intervals. *Behavior Research Methods and Instrumentation, 5*, 263–265.

Manning, S.K. (1973b). A measure of spontaneous alternation. *Behavior Research Methods and Instrumentation, 5*, 501–502.

Markowska, A., & Lukaszewska, I. (1974). Short-term memory of spatio-visual events preserved after frontomedial or frontopolar lesions in rats. *Acta Neurobiologica Experimentia, 34*, 715–721.

Marshall, J.F. (1979). Somatosensory inattention after dopamine depleting intracerebral 6-OHDA injections: Spontaneous recovery and pharmacological control. *Brain Research, 177*, 311–324.

Marshall, J.F., Richardson, J.S., & Teitelbaum, P. (1974). Nigrostriatal bundle damage and the lateral hypothalamic syndrome. *Journal of Comparative and Physiological Psychology, 87*, 808–830.

Marshall, J.F., & Teitelbaum, P. (1974). Further analysis of sensory inattention following lateral hypothalamic damage in rats. *Journal of Comparative and Physiological Psychology, 86*, 375–395.

Mason, S.T., Wood, C., & Angel, A. (1983). Brain noradrenaline and varieties of alternation learning. *Behavioral Brain Research, 9*, 119–127.

May, R.B., & Wellman, A.W. (1968). Alternation in the fruit fly, *Drosophila melanogaster. Psychonomic Science, 12*, 339–340.

McDonald, B.E. (1986). *Memory impairment following repeated organophosphate exposure*. Unpublished master's thesis (34294), University of Washington, Seattle.

McGuiness, C., Dalsass, M., Proshansky, E., & Krauthamer, G. (1976). Afferent connections of the nucleus centrum medianum in the cat. *Neuroscience Abstracts, 2*, 67.

McNaughton, B.L., & Feldon, J. (1980). Spontaneous alternation of body turns and place: Differential effects of amylobarbitone, scopolamine and septal lesions. *Psychopharmacology, 68*, 201–206.

McNaughton, B.L., & Morris, R.G.M. (1987). Hippocampal synaptic enhancement and information storage within a distributed memory system. *Trends in Neuroscience, 10*, 408–415.

Meador, K.J., Hammond, E.J., Loring, D.W., Allen, M., Bowers, D., & Heilman, K.M. (1987). Cognitive evoked potentials and disorders of recent memory. *Neurology, 37*, 526–529.

Meador, K.J., Loring, D.W., Adams, R.J., Patel, B.R., & Davis, H.C. (1987). Central cholinergic systems and the P3 evoked potential. *International Journal of Neuroscience, 33*, 199–205.

Meador, K.J., Loring, D.W., Bowers, D., & Heilman, K.M. (1987). Remote memory and the neglect syndrome. *Neurology, 37*, 522–526.

Means, L.W., Harrell, T.H., Mayo, E.S., & Alexander, G.B. (1974). Effects of dorsomedial thalamic lesions on spontaneous alternation, maze, activity and runway performance in the rat. *Physiology and Behavior, 12*, 973–979.

Means, L.W., Leander, J.D., & Isaacson, R.L. (1971). The effects of hippocampal lesions on alternation behavior and response to stimulus change. *Physiology and Behavior, 6*, 17–22.

Mellgren, R.L., & Hunsicker, J.P. (1975). Conditions of pre-exposure and passive avoidance behavior in rats. *Animal Learning & Behavior, 3*, 147–151.

Merkel, F.W. (1978). Angle sense in painted quails: A parameter of geodetic orientation. In K. Schmidt-Koenig & W.T. Keeton (Eds.), *Animal migration, navigation, and homing* (pp. 269–274). Berlin: Springer-Verlag.

Merkel, F.W., & Fischer-Klein, K. (1973). Winkelkompensation bei Zwergwachtein (*Excalfactoria chinensis*). *Die Vogelwarte, 27*, 39–50.

Mesulam, M.-M. (1981). A cortical network for directed attention and unilateral neglect. *Annals of Neurology, 10*, 309–325.

Meyers, B., & Domino, E.F. (1964). The effect of cholinergic blocking drugs on spontaneous alternation in rats. *Archives Internationales de Pharmacodynamie et de Therapie, 150*, 525–529.

Miller, F.P., Cox, R.H., Snodgrass, W.R., & Maickel, R.P. (1970). Comparative effects of p-chlorophenylalanine, p-chloroamphetamine and p-chloro-N-methylamphetamine on rat brain norepinephrine, serotonin, and 5-hydroxyindole-3-acetic acid. *Biochemistry and Pharmacology, 19*, 435–444.

Miller, J.S., Jagielo, J.A., & Spear, N.E. (in press). Age-related differences in short-term retention of separable elements of an odor aversion. *Journal of Experimental Psychology: Animal Behavior Processes.*

Miller, J.S., Nonneman, A.J., Kelly, K.S., Neisewander, J.L., & Isaac, W.L. (1986). Disruption of neophobia, conditioned odor aversion, and conditioned taste aversion in rats with hippocampal lesions. *Behavioral and Neural Biology, 45*, 240–253.

Miller, J.S., & Spear, N.E. (1989). Ontogenetic differences in short-term retention of Pavlovian conditioning. *Developmental Psychobiology, 22*, 377–387.

Miller, J.S., Potegal, M., & Abraham, L. (1983). Vestibular involvement in a passive transport and return task. *Physiological Psychology, 11*, 1–10.

Milner, B. (1985). Memory and the human brain. In M. Shafto (Ed.), *How we know* (pp. 31–59). San Francisco: Harper & Row.

Milner, B., Corkin, S., & Teuber, H.-L. (1968). Further analysis of the hippocampal amnesic syndrome: 14-year follow-up study of H.M. *Neuropsychologia, 6*, 215–234.

Milner, B., Petrides, M., & Smith, M.L. (1985). Frontal lobes and the temporal organization of memory. *Human Neurobiology, 4*, 137–142.

Mitchell, J.A., & Hall, G. (1987). Basal ganglia, instrumental and spatial learning. In P. Ellen & C. Thinus-Blanc (Eds.), *Cognitive processes and spatial orientation in animal and man* (Vol. 2, pp. 124–130). Boston: Martinus Nijhoff.

Mittelstaedt, M.L. (1972). Idiothetic course control and visual orientation. In R. Wehner (Ed.), *Information processing in the visual systems of arthropods* (pp. 275–279). Berlin: Springer-Verlag.

Mittelstaedt, M.L., & Mittelstaedt, H. (1980). Homing by path integration in a mammal. *Naturwissenschaften*, *67*, 566–567.

Mittelstaedt, M.L., Mittelstaedt, H., & Mohren, W. (1979). Interaction of gravity and idiothetic course control in millipedes. *Journal of Comparative Physiology*, *133*, 267–281.

Montgomery, K.C. (1951). Spontaneous alternation as a function of time between trials and amount of work. *Journal of Experimental Psychology*, *42*, 82–93.

Montgomery, K.C. (1952). A test of two explanations of spontaneous alternation. *Journal of Comparative and Physiological Psychology*, *45*, 287–293.

Montgomery, K.C. (1953). The effect of hunger and thirst drives upon exploratory behavior. *Journal of Comparative and Physiological Psychology*, *46*, 315–319.

Montone, K.T., Fass, B., & Hamill, G.S. (1988). Serotonergic and nonserotonergic projections from the rat interpeduncular nucleus to the septum, hippocampal formation and raphe: A combined immunocytochemical and fluorescent retrograde labelling study of neurons in the apical subnucleus. *Brain Research Bulletin*, *20*, 233–240.

Moore, F.R. (1982). Reverse turning and angle compensation in a migratory passerine, *Passerculus sandwichensis*. *Behavioural Processes*, *7*, 259–266.

Morley, B.J. (1986). The interpeduncular nucleus. *International Review of Neurobiology*, *28*, 157–182.

Morrow, L.A. (1987). Cerebral lesions and internal spatial representations. In P. Ellen & C. Thinus-Blanc (Eds.), *Cognitive processes and spatial orientation in animal and man* (Vol. 2, pp. 156–164). Boston: Martinus Nijhoff.

Motamura, N., Yamadori, A., Ogura, J., Sakai, T., & Sawada, T.(1986). Unilateral spatial neglect due to hemorrhage in the thalamic region. *Acta Neurologica Scandinavica*, *74*, 190–194.

Mowrer, O.H., & Jones, H.M. (1943). Extinction and behavior variability as functions of effortfulness of task. *Journal of Experimental Psychology*, *33*, 369–386.

Moye, T.B., & Rudy, J.W. (1987). Ontogenesis of trace conditioning in young rats: Dissociation of associative and memory processes. *Developmental Psychobiology*, *20*, 405–414.

Murphey, R.M. (1965). Sequential alternation behavior in the fruit fly, *Drosophila melanogaster*. *Journal of Comparative and Physiological Psychology*, *60*, 196–199.

Myhrer, T. (1975a). Effects of stria terminalis sections on locomotor, avoidance, and alternation behavior in rats. *Physiological Psychology*, *3*, 245–248.

Myhrer, T. (1975b). Locomotor, avoidance and maze behavior in rats with selective disruption of hippocampal output. *Journal of Comparative and Physiological Psychology*, *89*, 759–777.

Nauta, W.J.H. (1958). Hippocampal projections and related neural pathways to the midbrain in the cat. *Brain*, *81*, 319–340.

Nauta, W.J.H. (1960). Some neural pathways related to the limbic system. In E. Ramey & D. O'Doherty (Eds.), *Electrical studies on the unanesthetized brain* (pp. 1–17). New York: Paul B. Hoeber.

Nauta, W.J.H., & Domesick, V.B. (1984). Afferent and efferent relationships of the basal ganglia. In *Ciba Foundation Symposium 107* (pp. 3-23). London: Pitman.

Neafsey, E.J., Hull, C.D., & Buchwald, N.A. (1978). Preparation for movement in the cat. II. Unit activity in the basal ganglia and thalamus. *Electroencephalography and Clinical Neurophysiology*, *44*, 714–723.

Neiberg, A., Dale, J., & Grainger, D. (1970). Alternation of stimulus and response in three species. *Psychonomic Science*, *181*, 183–184.

Nissen, H.W. (1950). Description of the learned response in discrimination behavior. *Psychological Review, 57*, 121–131.

Nonneman, A.J., & Kolb, B. (1979). Functional recovery after serial ablation of prefrontal cortex in the rat. *Physiology and Behavior, 22*, 895–901.

Oades, R., Taghzouti, K., Simon, H., & LeMoal, M. (1985). Dopamine-sensitive alternation and collateral behavior in a Y-maze: Effects of d-amphetamine and haloperidol. *Psychopharmacology, 85*, 123–128.

O'Connell, R.H. (1964). Comparison of alternation and response to stimulus change. *Journal of Comparative and Physiological Psychology, 57*, 362–366.

O'Connell, R.H. (1971). Spontaneous alternation of brightness? *Psychonomic Science, 22*, 273–274.

O'Keefe, J., & Nadel, L. (1979). Precis of "The hippocampus as a cognitive map." *Behavioral and Brain Sciences, 2*, 487–533 (plus commentaries).

O'Keefe, J., & Speakman, A. (1987). Single unit activity in the rat hippocampus during a spatial memory task. *Experimental Brain Research, 68*, 1–27.

Oliverio, A. (1968). Effects of scopolamine on avoidance conditioning and habituation of mice. *Psychopharmacologia, 12*, 214–216.

Olton, D.S. (1983). Memory functions and the hippocampus. In W. Seifert (Ed.), *Neurobiology of the hippocampus (pp. 335*–373). New York: Academic Press.

Olton, D.S., Becker, J.T., & Handelmann, G.E. (1979). Hippocampus, space and memory. *Behavioral and Brain Sciences, 2*, 313–365. (plus commentaries).

Olton, D.S., Wenk, G.L., Church, R.M., & Meck, W.H. (1988). Attention and the frontal cortex as examined by simultaneous temporal processing. *Neuropsychologia, 26*, 307–318.

Overstreet, D.H. (1977). Pharmacological approaches to habituation of the acoustic startle response in rats. *Physiological Psychology, 5*, 230–238.

Overton, D.A. (1966). State-dependent learning produced by depressant and atropine-like drugs. *Psychopharmacologia, 10*, 6–31.

Papez, J.W. (1937). A proposed mechanism of emotion. *Archives of Neurology and Psychiatry, 38*, 725–743.

Parker, M.V., Miller, S.W., & Groves, P.M. (1974). Neuronal habituation and sensitization in the reticular formation of the rat. *Physiological Psychology, 2*, 464–470.

Parkes, M.W. (1965). An examination of central actions characteristic of scopolamine: Comparison of central and peripheral activity in scopolamine, atropine and some synthetic esters. *Psychopharmacology, 7*, 1–19.

Parsons, P.J., Fagan, T., & Spear, N.E. (1973). Short-term retention of habituation by infant, weanling, and adult rats. *Journal of Comparative and Physiological Psychology, 84*, 543–545.

Pate, J.L., & Bell, G.L. (1971). Alternation behavior of children in a cross-maze. *Psychonomic science, 23*, 431–432.

Pate, J.L., & DeLoache, J. (1970). Brightness and direction as cues for spontaneous alternation behavior. *Psychonomic Science, 18*, 27–38.

Payne, R., & Anderson, D.C. (1967). Scopolamine-produced changes in activity and in the startle response: Implications for behavioral activation. *Psychopharmacologia, 12*, 83–90.

Pellegrino, L.J., & Altman, J. (1979). Effects of differential interference with postnatal cerebellar neurogenesis on motor performance, activity level, and maze learning of rats: A developmental study. *Journal of Comparative and Physiological Psychology, 93*, 1–33.

Penfield, W. (1957). Vestibular sensation and the cerebral cortex. *Annals of Otology, 66*, 691–698.

Penfield, W., & Milner, B. (1958). Memory deficit produced by bilateral lesions in the hippocampal zone. *AMA Archives of Neurology and Psychiatry, 79*, 475–497.

Petchkovsky, L., & Kirkby, R.J. (1970). Individual differences, emotionality, and spontaneous alternation in mice. *Australian Journal of Psychology, 22*, 75–78.

Peterson, J. (1917). The effect of length of blind alleys on maze learning. *Behavior Monographs, 15*, 53.

Platt, J.J., & James, W.T. (1967). Response to stimulus change in the opossum. *Journal of Psychology, 67*, 85–89.

Platt, J.J., Yaksh, T., & James, W.T. (1968). Response to environmental stimulus change in young opossums. *Perceptual and Motor Skills, 26*, 43–46.

Pletscher, A., Bartholini, G., Bruderer, H., Bunkard, W.P., & Grey, K.F. (1964). Chlorinated aralkylamines affecting the cerebral metabolism of 5-hydroxytryptamine. *Journal of Pharmacology and Experimental Therapeutics, 145*, 344–353.

Pompeiano, O. (1974). Cerebellar-vestibular interrelations. In H. Kornhuber (Ed.), *Handbook of sensory physiology* (Vol. VI/1, pp. 417–476). Berlin: Springer-Verlag.

Potegal, M. (1969). The role of the caudate nucleus in spatial orientation of rats. *Journal of Comparative and Physiological Psychology, 69*, 756–764.

Potegal, M. (1982). Vestibular and neostriatal contributions to spatial orientation. In M. Potegal (Ed.), *Spatial abilities* (pp. 361–387). New York: Academic Press.

Potegal, M. (1987). The vestibular navigation hypothesis: A progress report. In P. Ellen & C. Thinus-Blanc (Eds.). *Cognitive processes and spatial orientation in animal and man* (pp. 29–34). Boston: Martinus Nijhoff.

Potegal, M., Copack, P., deJong, J., Krauthamer, G., & Gilman, S. (1971). Vestibular input to the caudate nucleus. *Experimental Neurology, 32*, 448–465.

Potegal, M., Day, M.J., & Abraham, L. (1977). Maze orientation, visual and vestibular cues in two-maze spontaneous alternation of rats. *Physiological Psychology, 5*, 414–420.

Potegal, M., & Squire, L. (1974). Contribution of the caudo-putamen to spontaneous alternation in rats. *Neuroscience Abstracts*, 1974, 378.

Prokasy, W.F. (1960). Postasymptotic performance decrements during massed reinforcements. *Psychological Bulletin, 57*, 237–247.

Pubols, B.H. (1956). The facilitation of visual and spatial discrimination reversal by overlearning. *Journal of Comparative and Physiological Psychology, 49*, 243–248.

Putnam, C.D. (1962). The non-random behaviour of *Aleochara bilineata* Gyll. (Coleoptera Staphylinidae) in a Y-maze with neither reward nor punishment in either arm. *Animal Behaviour, 10*, 118–125.

Ramirez, J.J., & Stein, D.G. (1984). Sparing and recovery of spatial alternation performance after entorhinal cortex lesions in rats. *Behavioral Brain Research, 13*, 53–61.

Randrup, A., & Munkvad, I. (1970). Biochemical, anatomical and psychological investigations of stereotyped behavior induced by amphetamines. In E. Costa and S. Garattini (Eds.), *Amphetamines and related compounds* (pp. 695–714). New York: Raven Press.

Ratcliff, G., & Newcombe, F. (1973). Spatial orientation in man: Effects of left, right, and bilateral posterior cerebral lesions. *Journal of Neurology, Neurosurgery and Psychiatry, 36*, 448–454.

Reeves, A.G., & Plum, F. (1969). Hyperphagia, rage, and dementia accompanying a ventromedial hypothalamic neoplasm. *Archives of Neurology, 20*, 616–623.

Restle, F. (1957). Discrimination of cues in mazes: A resolution of the "place-vs.-response" question. *Psychological Review, 64*, 217–228.

Rice, G.E., & Lawless, R.H. (1957). Behavior variability and reactive inhibition in the maze behavior of *Planaria dorotocephala. Journal of Comparative and Physiological Psychology, 50*, 105–108.

Richman, C.L. (1969). *Drive and reward effects on spontaneous alternation*. Unpublished manuscript. Wake Forest University, Department of Psychology, Winston-Salem, NC.

Richman, C.L. (1970) *Effects of overtesting, drive and reward on spontaneous alternation*. Unpublished manuscript. Wake Forest University, Department of Psychology, Winston-Salem, NC.

Richman, C.L., Chappell, L.R., Crowe, D., Perry, P.C., & Redfearn, L. (1970). Spontaneous alternation behavior in rats as a function of brightness cues. *Psychological Reports, 27*, 356-358.

Richman, C.L., & Coussens, W. (1970). Undertraining reversal effect in rats. *Journal of Experimental Psychology, 86*, 340–342.

Richman, C.L., Dember, W.N., & Kim, P. (1987). Spontaneous alternation behavior in animals: A review. *Current Psychological Research & Reviews, 5*, 358–391.

Richman, C.L., Gardner, J.T., Montgomery, M.D., & Benewicz, K.L. (1970). Effects of body weight loss on position and brightness discrimination tasks. *Learning and Motivation, 1*, 218–225.

Richman, C.L., Gulkin, R., & Knoblock, K. (1972). Effects of bulbectomization, strain, and gentling on emotionality and exploratory behavior in rats. *Physiology and Behavior, 8*, 447–452.

Richman, C.L., Knoblock, K., & Coussens, W. (1972). The overtraining reversal effect in rats: A function of task difficulty. *Quarterly Journal of Experimental Psychology, 24*, 291–298.

Ridley, R.M., Baker, H.F., & Haystead, T.A.J. (1981). Perseverative behavior after amphetamine: Dissociation of response tendency from reward association. *Psychopharmacology, 75*, 283–286.

Riley, D.A., & Shapiro, A.N. (1952). Alternation behavior as a function of effortfulness of task and distribution of trials. *Journal of Comparative and Physiological Psychology, 45*, 468–475.

Roberts, W.W., Dember, W.N., & Brodwick, M. (1962). Alternation and exploration in rats with hippocampal lesions. *Journal of Comparative and Physiological Psychology, 55*, 695–700.

Robinson, E.S. (1934). Work of the integrated organism. In C. Murchison (Ed.), *Handbook of general experimental psychology* (pp. 571–650). Worcester, MA: Clark University Press.

Rosen, J.J., & Stein, D.G. (1968). *The role of vestibular and olfactory cues and the effects of limbic-rhinal system lesions on spontaneous alternation*. Unpublished manuscript.

Rosenstock, J., Field, T., & Greene E. (1977). The role of the mammillary bodies in spatial memory. *Experimental Neurology, 55*, 340–352.

Rubens, A.B. (1985). Caloric stimulation and unilateral visual neglect. *Neurology, 35*, 1019–1024.

Russell, E.M., & Pearce, G.A. (1971). Exploration of novel objects by marsupials. *Behaviour, 401*, 312–322.

Sachs, L.B., Klopfer, F.D., & Morrow, J.E. (1965). "Reactive inhibition" in the sowbug. *Psychological Reports, 17*, 739-743.

Salamone, J.D., Beart, P.M., Alpert, J.E., & Iverson, S.D. (1984). Impairment in T-maze alternation performance following nucleus basalis magnocellularis lesions in rats. *Behavioral Brain Research, 13*, 63-70.

Sanders-Bush, E., Bushing, J.A., & Sulser, F. (1975). Long-term effects of p-chloroamphetamine and related drugs on central serotonergic mechanisms. *Journal of Pharmacology and Experimental Therapeutics, 192*, 33-42.

Saporta, S., & Greene, E. (1974). Orienting bias in the rat produced by hippocampal lesions. *Behavioral Biology, 11*, 339-351.

Sarter, M., Bodewitz, G., & Stephens, D.N. (1968). Attenuation of scopolamine-induced impairment of spontaneous alternation behavior by antagonist but not inverse agonist and agonist beta-carbolines. *Psychopharmacology, 94*, 491-495.

Schäfer, M.W. (1972). Reverse turning in Lithobius forficatus L. *Monitore Zoologico Italiano, 6*, 179-194.

Schäfer, M.W. (1975a). Gegendrehung und Winkelsinn in der Orientierung von *Lithobius forficatus* L. Teil I: Nachweis des Winkelsinnes, Charakterisierung der Orientierungsleistung. *Behaviour, 55*, 15-41.

Schäfer, M.W. (1975b). Gegendrehung und Winkelsinn in der Orientierung von *Lithobius forficatus* L. Teil II: Die auberen und inneren Bezugspunkte. *Behaviour, 55*, 42-72.

Schäfer, M.W. (1976a). Aspects of spontaneous orientation and bias in the maze-behaviour of Lithobius forficatus L. (Myriapoda Chilopoda) and the white mouse (Mus musculus L.). *Monitore Zoologicao Italiano, 10*, 205-217.

Schäfer, M.W. (1976b). Thigmotactic behaviour in Lithobius forficatus L. (Myriapoda Chilopoda). *Monitore Zoologico Italiano, 10*, 191-204.

Schäfer, M.W. (1980). Gegendrehung und Winkelsinn in der Orientierung von *Lithobius forficatus* L. Teil III: Erganzende Untersuchungen zur Verrechnung von Winkelfolgen. *Behaviour, 73*, 106-129.

Schäfer, M.W. (1982a). Ein idiothetischer Mechanismus im Gegendrehungsverhalten der Assel *Oniscus* L. *Zoologische Jahrbucher. Abteilung für Allgemeiene Zoologie und Physiologie der Tiere, 86*, 193-208.

Schäfer, M.W. (1982b). Gegendrehung und Winkelsinn in der Orientierung verschiedener Arthropoden. *Zoologische Jahrbucher. Abteilung für Allgemeine Zoologie und Physiologie der Tiere, 86*, 1-16.

Schäfer, M.W. (1983). Analyse des Labyrinthverhaltens von *Lithobius forficatus* L. (Myriapoda, Chilopoda). *Zoologische Jahrbucher. Abteilung für Allgemeine Zoologie und Physiologie der Tiere, 87*, 141-172.

Schäfer, M.W. (1986). Erganzende Untersuchungen zum Gegendrehungsverhalten der Assein *Oniscus asellus* L. und *Porcellio scaber* LATR. *Zoologische Jahrbucher. Abeilung für Allgemeine Zoologie und Physiologie der Tiere, 90*, 285-296.

Schäfer, M.W., & Wegner, U. (1974). Reverse turning and angle sense in Tenebrio molitor L. *Monitore Zoologica Italiano, 8*, 141-149.

Scheaer, C., & Moore, J.N. (1974). Breakdown of stimulus control by d-amphetamine. *Physiological Psychology, 2*, 427-430.

Schechter, M.D., & Cook, P.G. (1975). Dopaminergic mediation of the interoceptive cue produced by d-amphetamine in rats. *Psychopharmacologia, 42*, 185-193.

Schiorring, E. (1979). An open field study of stereotyped locomotor activity in amphetamine treated rats. *Psychopharmacology, 66*, 281-287.

Schneirla, T.C. (1929). Learning and orientation in ants. *Comparative Psychology Monographs, 6*, 1-142.

Schnurr, R. (1971). Spontaneous alternation in normal and brain-damaged gerbils. *Psychonomic Science, 25*, 181-182.

Schultz, D.P. (1964). Spontaneous alternation behavior in humans: Implications for psychological research. *Psychological Bulletin, 63*, 394-400.

Schwaber, J.S., Rogers, W.T., Satoh, K., & Fibiger, H.C. (1987). Distribution and organization of cholinergic neurons in the rat forebrain demonstrated by computer-aided data acquisition and three-dimensional reconstruction. *Journal of Comparative Neurology, 263*, 309-325.

Semba, K., Reiner, P.B., McGeer, E.G., & Fibiger, H.C. (1988). Brainstem afferents to the magnocellular basal forebrain studied by axonal transport, immunohistochemistry, and electrophysiology in the rat. *Journal of Comparative Neurology, 267*, 433-453.

Semmes, J., Weinstein, S., Ghent, L., & Teuber, H.-L. (1955). Spatial orientation in man after cerebral injury: 1. Analysis by locus of lesion. *Journal of Psychology, 39*, 227-244.

Shepard, J.F. (1959). *An unexpected cue in maze learning*. Ann Arbor: Edwards Brothers.

Sherrick, M.F., Brunner, R.L., Roth, T.G., & Dember, W.N. (1979). Rats' sensitivity to their direction of movement and spontaneous alternation behavior. *Quarterly Journal of Experimental Psychology, 31*, 83-93.

Sherrick, M.F., & Dember, W.N. (1966a). The tendency to alternate direction of movement as reflected in starting stem running speed. *Psychonomic Science, 6*, 29-30.

Sherrick, M.F., & Dember, W.N. (1966b). Trial-two goal arm alternation to direction of movement in trial-one straight alley. *Psychonomic Science, 6*, 317-318.

Sherrington, C.S. (1906). *The integrative action of the nervous system*. New Haven, CT: Yale University Press.

Shinkman, P.G., & Hertzler, D.R. (1964). Maze alternation in the planarian, *Dugesia tigrina*. *Psychonomic Science, 1*, 407-408.

Sidman, M., Stoddard, L.T., & Mohr, J.P. (1968). Some additional quantitative observations of immediate memory in a patient with bilateral hippocampal lesions. *Neuropsychologia, 6*, 245-254.

Silberpfennig, J. (1941). Contributions to the problem of eye movements. *Confinia Neurologica, 4*, 1-13.

Sinclair, J.D., & Bender, D.O. (1978). Compensatory behaviors: Suggestion for a common basis from deficits in hamsters. *Life Sciences, 22*, 1407-1412.

Singh, D., (1973). Comparison of behavioral deficits caused by lesions in septal and ventromedial hypothalamic areas of female rats. *Journal of Comparative and Physiological Psychology, 84*, 370-379.

Sinha, S.N., Francks, C.M., & Broadhurst, P.L. (1958). The effect of a stimulant and a depressant drug on a measure of reactive inhibition. *Journal of Experimental Psychology, 56*, 349-354.

Skinner, B.F. (1942). The processes involved in the repeated guessing of alternatives. *Journal of Experimental Psychology, 30*, 495-503.

Skinner, B.F. (1948). "Superstitution" in the pigeon. *Journal of Experimental Psychology, 381*, 168-172.

Smith, G.J., Goodlett, C.R., Burright, R.G., Donovick, P.J., & Spear, N.E. (1983). The presence of home-cage stimuli attenuates spontaneous-alternation deficits in rats with septal lesions. *Physiological Psychology, 11*, 119-124.

Smith, G.J., Kucharski, D., & Spear, N.E. (1985). Conditioning of an odor aversion in preweanlings with isolation from home nest as the unconditioned stimulus. *Developmental Psychobiology, 18*, 421–434.

Smith, G.J., & Spear, N.E. (1978). Effects of the home environment on withholding behaviors and conditioning in infant and neonatal rats. *Science, 202*, 327–329.

Smith, M.L. (1987). The encoding and recall of spatial location after right hippocampal lesions in man. In P. Ellen & C. Thinus-Blanc (Eds.), *Cognitive processes and spatial orientation in animal and man* (Vol. 2, pp. 165–172). Boston: Martinus Nijhoff.

Smith, M.L., & Milner, B. (1988). Estimation of frequency of occurrence of abstract designs after frontal or temporal lobectomy. *Neuropsychologia, 26*, 297–306.

Solomon, P.R., & Moore, J.W. (1975). Latent inhibition and stimulus generalization of the classically conditioned nictitating membrane response in rabbits following dorsal hippocampal ablation. *Journal of Comparative and Physiological Psychology, 89*, 1192–1203.

Solomon, R.L. (1948). The influence of work on behavior. *Psychological Bulletin, 45*, 1–40.

Sorenson, C.A., & Davis, M. (1975). The effects of 6-hydroxydopamine and alpha-methyl-para-tyrosine on the acoustic startle in rats. *Pharmacology, Biochemistry and Behavior, 3*, 325–329.

Sorensen, K.E., & Witter, M.P. (1983). Entorhinal efferents reach the caudatoputamen. *Neuroscience Letters, 35*, 259–264.

Spear, N.E. (1978). *The processing of memories: Forgetting and retention.* Hillsdale, NJ: Erlbaum.

Spear, N.E. (1979). Experimental analysis of infantile amnesia. In J.F. Kihlstrom & F.J. Evans (Eds.), *Functional disorders of memory* (pp. 75–101). Hillsdale, NJ: Erlbaum.

Spear, N.E. (1984). Ecologically determined dispositions control the ontogeny of learning and memory. In R. Kail & N.E. Spear (Eds.), *Comparative perspectives on the development of memory* (pp. 325–357). Hillsdale, NJ: Erlbaum.

Spear, N.E., & Campbell, B.A. (1979). *Ontogeny of learning and memory.* Hillsdale, NJ: Erlbaum.

Spear, N.E., & Rudy, J. (in press). Tests of learning and memory in the developing rat. In H.N. Shair, G.A. Barr, & M.A. Hofer (Eds.), *Developmental psychobiology; Current methodological and conceptual issues.* New York: Oxford University Press.

Speedie, L.J., & Heilman, K.M. (1982). Amnesic disturbance following infarction of the left dorsomedial nucleus of the thalamus. *Neuropsychologia, 20*, 597–604.

Speedie, L.J., & Heilman, K.M. (1983). Anterograde memory deficits for visuospatial material after infarction of the right thalamus. *Archives of Neurology, 40*, 183–186.

Sperling, S.E. (1970). The ORE in simultaneous and differential reversal: Acquisition tasks, acquisition criterion, and reversal task. *Journal of Experimental Psychology, 84*, 349–360.

Spigel, I.M., & Ellis, K. (1965). An emergent albedo preference in turtle. *Perceptual and Motor Skills, 21*, 113–114.

Squire, L.R. (1969). Effects of pretrial and posttrial administration of cholinergic and anticholinergic drugs on spontaneous alternation. *Journal of Comparative and Physiological Psychology, 69*, 69–75.

Squire, L.R., & Moore, R.Y. (1979). Dorsal thalamic lesions in a noted case of human memory dysfunction. *Annals of Neurology, 6*, 503–506.

Staddon, J.E.R., & Simmelhag, V. (1971). The "superstition" experiment: A reexamination of its implications for the principles of adaptive behavior. *Psychological Review, 78*, 3–43.

Stein, L. (1966). Habituation and stimulus novelty: A model based on classical conditioning. *Psychological Review, 73*, 352–356.

Stellar, E. (1954). The physiology of emotion. *Psychological Review, 61*, 5–22.

Stevens, R. (1973). Effects of duration of sensory input and intertrial interval on spontaneous alternation in rats with hippocampal lesions. *Physiological Psychology, 1*, 41–44.

Stevens, R., & Cowey, A. (1973). Effects of dorsal and ventral hippocampal lesions on spontaneous alternation, learned alternation and probability learning in rats. *Brain Research, 52*, 203–224.

Steward, O., Loesche, J., & Horton, W.C. (1977). Behavioral correlates of denervation and reinnervation of the hippocampal formation of the rat: Open field activity and cue utilization following bilateral entorhinal cortex lesions. *Brain Research Bulletin, 2*, 41–48.

Stewart, W.J., & Blain, S. (1975). Dose-response effects of scopolamine on activity in an open-field. *Psychopharmacologia, 44*, 291–295.

Still, A., & MacMillan, A. (1969). Odor trail and spontaneous alternation. *Psychonomic Science, 16*, 160–161.

Storm-Mathisen, J., & Goldberg, H.C. (1974). 5-Hydroxytryptamine and noradrenaline in the hippocampal region: Effects of transection of afferent pathways on endogenous levels, high affinity uptake and some transmitter related enzymes. *Journal of Neurochemistry, 22*, 793–803.

Stumpf, C. (1965). Drug action on the electrical activity of the hippocampus. *International Review of Neurobiology, 8*, 77–138.

Stuss, D.R., & Benson, D.F. (1983). Frontal lobe lesions and behavior. In A. Kertesz (Ed.), *Localization in neuropsychology* (pp. 429–454). New York: Academic Press.

Sutherland, N.S. (1957). Spontaneous alternation and stimulus avoidance. *Journal of Comparative and Physiological Psychology, 50*, 358–362.

Sutherland, N.S., Mackintosh, N.J., & Mackintosh, J. (1965). Shape and size discrimination in *Octopus*: The effects of pretraining along different dimensions. *Journal of Genetic Psychology, 106*, 1–10.

Svensson, T., Bunney, B., & Aghajanian, G. (1975). Inhibition of both noradrenergic and serotonergic neurons in brain by the alpha-adrenergic agonist clonidine. *Brain Research, 92*, 291–306.

Swanson, L.W. (1981). A direct projection from Ammon's horn to prefrontal cortex in the rat. *Brain Research, 217*, 150–154.

Swonger, A.K., Chambers, W., & Rech, R.H. (1970). The effects of alteration in brain 5-HT on habituation of the cortical evoked response and the startle response in rats. *Pharmacologist, 12*, 207–214.

Swonger, A.K., & Rech, R.H. (1972). Serotonergic and cholinergic involvement in habituation of activity and spontaneous alternation of rats in a Y-maze. *Journal of Comparative and Physiological Psychology, 81*, 509–522.

Swonger, A.K., & Rech, R.H. (1977). Hallucinogens. In H. Anisman and G. Bignami (Eds.), *Psychopharmacology of aversively motivated behaviors* (pp. 345–383). New York: Plenum Press.

Syme, G.J., & Syme, L.A. (1977). Spontaneous alternation in mice: A test of the mere-exposure hypothesis. *American Journal of Psychology, 90*, 621–633.

Taghzouti, K., Louilot, A., Herman, J.P., LeMoal, M., & Simon, H. (1985). Alternation behavior, spatial discrimination, and reversal disturbances following 6-hydroxydopamine lesions in the nucleus accumbens of the rat. *Behavioral and Neural Biology, 34*, 98-103.

Taghzouti, K., Simon, H., & LeMoal, M. (1986). Disturbances in exploratory behaviour and functional recovery in the Y and radial mazes following dopamine depletion of the lateral septum. *Behavioral and Neural Biology, 45*, 48-56.

Talland, G.A. (1968). *Disorders of memory and learning.* Baltimore: Penguin.

Telford, C.W. (1931). The refractory phase of voluntary and associative responses. *Journal of Experimental Psychology, 14*, 1-36.

Teyler, T.J., & DiScenna, P. (1986). The hippocampal memory indexing theory. *Behavioral Neuroscience, 100*, 147-154.

Thomas, G.J. (1979). Comparison of effects of small lesions in posterodorsal septum on spontaneous and rerun correction (contingently reinforced) alternation in rats. *Journal of Comparative and Physiological Psychology, 93*, 685-696.

Thomas, R.K., & Weir, V.K. (1975). The effects of lesions in the frontal or posterior association cortex of rats on maze III. *Physiological Psychology, 3*, 210-214.

Thompson, M.E. (1960). Alternation in a T maze as a function of three variables. *Psychological Reports, 7*, 103-110.

Thompson, M.E., Estell, M., & Loomis, L.A. (1965). Response variability in the quail (Coturnix coturnix Japonica). *Psychonomic Science, 3*, 519-520.

Thompson, R.F., Groves, P.M., Teyler, T.J., & Roemer, R.A. (1973). A dual-process theory of habituation: Theory and behavior. In H.V.S. Peeke and M.J. Herz (Eds.), *Habituation: Behavior studies* (I, 239-273). New York: Academic Press.

Thompson, R.F., & Lippman, L.G. (1972). Exploration and activity in the gerbil and rat. *Journal of Comparative and Physiological Psychology, 80*, 439-448.

Thorndike, E.L. (1927). The refractory period in associative processes. *Psychological Review, 34*, 234-236.

Thorpe, W.H. (1963). *Learning and instinct in animals.* London: Methuen.

Tigner, J.C. (1974). The effects of dorsomedial thalamic lesions on learning, reversal, and alternation behavior in the rat. *Physiology and Behavior, 12*, 13-17.

Tilley, M.W., Doolittle, J.N., & Mason, D.J. (1966). Spontaneous alternation in the Virginia opossum. *Psychological Reports, 19*, 593-594.

Tobe, A., Egawa, M., & Nagai, R. (1983). Effect of 4-(0-benzylphenoxyl)-n-methylbutylamine hydrochloride (MC-2016) on the scopolamine-induced deficit of spontaneous alternation behavior in rats. *Japanese Journal of Pharmacology, 33*, 775-784.

Tolman, E.C. (1925). Purpose and cognition: The determiners of animal learning. *Psychological Review, 32*, 285-297.

Tolman, E.C. (1932). *Purposive behavior in animals and men.* New York: The Century Co.

Tolman, E.C., Ritchie, B.F., & Kalish, D. (1946). Studies in spatial learning, I. Orientation and the short-cut. *Journal of Experimental Psychology, 36*, 13-24.

Urca, A.G., & Lubow, R.E. (1975). The effects of pre-exposure to a non-attended stimulus on subsequent learning: Latent inhibition in adults. *Bulletin of the Psychonomic Society, 5*, 5-8.

Vallar, G., & Perani, D. (1986). The anatomy of unilateral neglect after right hemisphere stroke lesions: A clinical/CT scan correlation study in man. *Neuropsychologia, 24*, 609-622.

Vanderwolf, C.H., Kramis, R., Gillespie, L.A., & Bland, B.H. (1975). Hippocampal rhythmic slow activity and neocortical low-voltage fast activity: Relations to behavior. In R. Isaacson & K. Pribram (Eds.), *The hippocampus* (Vol. 2, pp. 101–128). New York: Plenum Press.

Vanegas, H., & Flynn, J.P. (1968). Inhibition of cortically-elicited movement by electrical stimulation of the hippocampus. *Brain Research*, *11*, 489–506.

Van Hoesen, G.W. (1985). Neural systems of the non-human primate forebrain implicated in memory. *Annals of the New York Academy of Science*, *444*, 97–112.

Veening, J.G., Cornelissen, F.M., & Lieven, P.A.J.M. (1980). The topical organization of the afferents to the caudatoputamen of the rat: A horseradish peroxidase study. *Neuroscience*, *5*, 1253–1268.

Victor, M., Adams, R.D., & Collins, G.H. (1971). *The Wernicke-Korsakoff syndrome*. Philadelphia: F.A. Davis.

Villablanca, J.R., Olmstead, C.E., Levine, M.S. & Marcus, R.J. (1978). Effects of caudate nuclei or frontal cortical ablations in kittens: Neurology and gross behavior. *Experimental Neurology*, *61*, 615–634.

Vincent, S.B. (1915). The white rat and the maze problem, I. Introduction of visual control. *Journal of Animal Behavior*, *5*, 1–24.

Walk, R.D. (1960). Response of dark- and light-reared rats to stimulus change. *Journal of Comparative and Physiological Psychology*, *53*, 609–611.

Walker, E.L. (1956). The course and duration of the reaction decrement and the influence of reward. *Journal of Comparative and Physiological Psychology*, *49*, 167–176.

Walker, E.L. (1958). Action decrement and its relation to learning. *Psychological Review*, *65*, 129–142.

Walker, E.L., Dember, W.N., Earl, R.W., Fawl, C., & Karoly, A.J. (1955). Choice alternation III. Response intensity versus response discriminability. *Journal of Comparative and Physiological Psychology*, *48*, 80–85.

Walker, E.L., Dember, W.N., Earl, R.W., Fliege, S.W., & Karoly, A.J. (1955). Choice alternation II. Exposure to stimulus or stimulus and place without choice. *Journal of Comparative and Physiological Psychology*, *48*, 24–28.

Walker, E.L., Dember, W.N., Earl, R.W., & Karoly, A.J. (1955). Choice alternation I. Stimulus vs. place vs. response. *Journal of Comparative and Physiological Psychology*, *48*, 19–23.

Walters, A., & Block, R.G. (1969). Scopolamine effects on locomotor and exploratory activity in rats. *Psychonomic Science*, *17*, 3–4.

Warburton, D.M. (1972). The cholinergic control of internal inhibition. In R. Boakes & M. Halliday (Eds.), *Inhibition and learning* (pp. 431–460). London: Academic Press.

Watanabe, M., & Iwata, K. (1956). Alternative turning response of *Armadillidium vulgare*. *Annual of Animal Psychology*, *6*, 75–82.

Wayner, M.J., & Zellner, D.K. (1958). The role of the suprapharyngeal ganglion in spontaneous alternation and negative movements in *Lumbricus terrestris L. Journal of Comparative and Physiological Psychology*, *151*, 282–287.

Weis, B.J., & Means, L.W. (1980). A comparison of the effects of medial frontal, dorsomedial thalamic, and combination lesions on discrimination and spontaneous alternation in the rat. *Physiological Psychology*, *8*, 325–329.

Weiskrantz, L., & Warrington, E.K. (1975). The problem of the amnesic syndrome in man and animals. In R. Isaacson & K. Pribram (Eds.), *The hippocampus*. (Vol. 2, pp. 411–428). New York: Plenum Press.

Weiss, K.R., Friedman R., & McGregor, S. (1974). Effects of septal lesions on latent inhibition and habituation of the orienting response in rats. *Acta Neurobiologiae Experimentalis, 34*, 491–504.

Wenk, G., Hughley, D., Boundy, V., Kim, A., Walker, L., & Olton, D. (1987). Neurotransmitters and memory: Role of cholinergic, serotonergic, and noradrenergic systems. *Behavioral Neuroscience, 101*, 325–332.

Westlake, G.F., & Kleerekoper, H. (1970). Evidence for a memory process in the turning behavior of free swimming goldfish. *Canadian Journal of Zoology, 4*, 813–815.

White, S.R. (1974). Atropine, scopolamine and hippocampal lesion effects on alternation performance of rats. *Pharmacology, Biochemistry and Behavior, 2*, 297–307.

Will, B., Duluzarche, F., & Kelche, C. (1983). Does post-operative environment attenuate or exacerbate symptoms which follow hippocampal lesions in rats? *Behavioral Brain Research, 7*, 125–132.

Williams, J.M., Hamilton, L.W., & Carlton, P.L. (1974). Pharmacological and anatomical dissociation of two types of habituation. *Journal of Comparative and Physiological Psychology, 87*, 724–732.

Williams, J.M., Hamilton, L.W., & Carlton, P.L. (1975). Ontogenetic dissociation of two classes of habituation. *Journal of Comparative and Physiological Psychology, 89*, 733–737.

Wilson, D.M., & Hoy, R.R. (1968). Optomotor reaction, locomotory bias, and reactive inhibition in the milkweed bug *Oncopeltus* and the beetle *Zophabas. Zeitschrift für Vergleichende Physiologie, 58*, 136–152.

Wilson, M.M., & Fowler, H. (1976). Variables affecting alternation behavior in the cockroach, *blata orientalis. Animal Learning & Behavior, 4*, 490–494.

Wingfield, R.C. (1943). Some factors influencing spontaneous alternation in human subjects. *Journal of Comparative and Physiological Psychology, 35*, 237–243.

Winocur, G., Oxbury, S., Roberts, R., Agnetti, V., & Davis, C. (1984). Amnesia in a patient with bilateral lesions to the thalamus. *Neuropsychologia, 22*, 123–143.

Wirtshafter, D. (1981). The role of interpenduncular connections with the tegmentum in avoidance learning. *Physiology and Behavior, 26*, 985–989.

Wise, R.A. (1978). Catecholamine theories of reward: A critical review. *Brain Research, 152*, 215–247.

Witkin, H.A., & Schneirla, T.C. (1937). Initial maze behavior as a function of maze design. *Journal of Comparative Psychology, 23*, 275–304.

Woodruff, M.L., Baisden, R.H., & Douglas, J.R. (1981). Effect of cingulate and fornix lesions on emotional behavior in rabbits (*oryctolagus cuniculus*). *Experimental Neurology, 74*, 379–395.

Woods, P.J., & Jennings, S. (1959). Response to environmental change: A further confirmation. *Psychological Reports, 5*, 560.

Worden, F.A. (1973). Auditory habituation. In H.V.S. Peeke and M.J. Herz (Eds.), *Habitutation: Physiological substrates*. (II, pp. 109–103). New York: Academic Press.

Wurtz, R.H., & Hikosaka, O. (1986). Role of the basal ganglia in the initiation of saccadic eye movements. *Progress in Brain Research, 64*, 175–190.

Yang, C.R., & Mogenson, G.J. (1984). Electrophysiological responses of neurones in the nucleus accumbens to hippocampal stimulation and the attenuation of the excitatory responses by the mesolimbic system. *Brain Research, 324*, 69–84.

Yunger, L.M., & Harvey, J.A. (1973). Effect of lesions in the medial forebrain bundle on three measures of pain-sensitivity and noise-elicited startle. *Journal of Comparative and Physiological Psychology, 83*, 173–181.

Zahner, H., Battig, K., & Grandjean, E. (1961). Das spontane Links-Rechts alternieren der Ratte und dessen einflussung durch psycopharmaka. *Helvetica Physiologia et Pharmacologia Acta, 19*, C38–39.

Zamora, E.N., & Kaelbling, R. (1965). Memory and electroconvulsive therapy. *American Journal of Psychiatry, 122*, 546–554.

Zeaman, D., & Angell, D. (1953). A spatial gradient of alternation tendency. *Journal of Comparative and Physiological Psychology, 46*, 390–392.

Zeaman, D., & House, B.J. (1951). The growth and decay of reactive inhibition as measured by alternation behavior. *Journal of Experimental Psychology, 41*, 177–186.

Zimbardo, P.G., & Montgomery, K.C. (1957). The relative strengths of consummatory responses in hunger, thirst and exploratory drive. *Journal of Comparative and Physiological Psychology, 50*, 504–508.

Zola-Morgan, S., & Squire, L.R. (1985). Amnesia in monkeys after lesions of the mediodorsal nucleus of the thalamus. *Annals of Neurology, 17*, 558–564.

Zola-Morgan, S., Squire, L.R., & Amalal, D.G. (1986). Human amnesia and the medial temporal region: Enduring memory impairment following a bilateral lesion limited to field CA1 of the hippocampus. *Journal of Neuroscience, 6*, 2950–2967.

# Author Index

# Subject Index

MIX
Papier aus verantwortungsvollen Quellen
Paper from responsible sources
FSC® C105338

If you have any concerns about our products,
you can contact us on
**ProductSafety@springernature.com**

In case Publisher is established outside the EU,
the EU authorized representative is:
**Springer Nature Customer Service Center GmbH**
**Europaplatz 3, 69115 Heidelberg, Germany**

Printed by Libri Plureos GmbH
in Hamburg, Germany